Biogenic Nanoparticles of Elemental Selenium: Synthesis, Characterization and Relevance in Wastewater Treatment

Thesis committee:

Thesis Promotor

Prof. dr. ir P.N.L. Lens
Professor of Biotechnology
UNESCO-IHE
Delft, The Netherlands

Thesis Co-Promotors

Dr. Hab. E.D. van Hullebusch,
Hab. Associate Professor in Biogeochemistry
University of Paris-Est
Marne-la-Vallée, France

Prof. François Farges
Professeur of minerology
National Museum of Natural History
Paris, France

Dr. G. Esposito,
Assistant Professor of Sanitary and Environmental Engineering
University of Cassino and Southern Lazio
Cassino, Italy

Other Members

Dr. E. Şahinkaya
Department of Bioengineering
Istanbul Medeniyet University
Istanbul, Turkey

Dr. Paul Mason
Department of Geosciences
Utrecht University
Utrecht, The Netherlands

This research was conducted under the auspices of the Erasmus Mundus Joint Doctorate Environmental Technologies for Contaminated Solids, Soils, and Sediments (ETeCoS3) and the Graduate School for Socio-Economic and Natural Sciences of the Environment (SENSE).

Joint PhD degree in Environmental Technology

UNIVERSITÉ
—PARIS-EST

Docteur de l'Université Paris-Est
Spécialité : Science et Technique de l'Environnement

Dottore di Ricerca in Tecnologie Ambientali

UNESCO-IHE
Institute for Water Education

Degree of Doctor in Environmental Technology

Thèse – Tesi di Dottorato – PhD thesis

Rohan JAIN

Biogenic nanoparticles of elemental selenium: synthesis, characterization and relevance in wastewater treatment

Defended on December 19th, 2014

In front of the PhD committee

Prof. Erkan Sahinkaya	Reviewer
Dr. Hab. Paul Mason	Reviewer
Prof. Piet Lens	Promotor
Prof. Stéphanie Rossano	Promotor
Dr. Hab. Eric van Hullebusch	Co-promotor
Dr. Hab. Giovanni Esposito	Co-promotor
Prof. Stefan Uhlenbrook	Examiner

European Commission
ERASMUS
MUNDUS

Erasmus Joint doctorate programme in Environmental Technology for Contaminated Solids, Soils and Sediments (ETeCoS3)

CRC Press/Balkema is an imprint of the Taylor & Francis Group, an informa business

© 2015, Rohan Jain

Published by:
CRC Press/Balkema
PO Box 11320, 2301 EH Leiden, The Netherlands
e-mail: Pub.NL@taylorandfrancis.com
www.crcpress.com – www.taylorandfrancis.com

ISBN 978-1-138-02831-9 (Taylor & Francis Group)

Table of contents   **Page no.**

Acknowledgment

It is my great pleasure to write this part of the thesis. This section would be certainly be a very long one as many people have contributed to put this thesis to this stage and it is my one of the chance to reciprocate. To begin with, I would like to thank Erasmus Mundus ETeCoS^3 program (FPA no. 2010-0009) for providing the financial support to carry out this thesis. I would also like to thank UNESCO-IHE, University of Paris-Est, University of Cassino and Southern Lazio, and National Museum of Natural History for hosting me.

I would like to sincerly thank Prof. Piet Lens, whose support, encouragement, push, scientific discussions and critical overview of the work was critical to achieve success. I also learned a lot from him and I am sure that this knowledge would serve me throughout my scientific career. I would like to show my gratitude toward Dr. Eric van Hullebusch whose constant inputs in providing ideas for the use of state-of-art techniques has been instrumental in improving the quality of the thesis. I would also like to acknowledge his support during the data collection at ESFR, Grenoble, without which, the successful data collection at ESRF would not have been possible. I would also like to thanks Prof. François Farges for introducing me to the world of theoritical calculation and XAS data analysis. I would like to acknowledge his patience in teaching basic principles of X-ray spectroscopy to a layman! I would also like to thank him for showing me some of the very fine restaurants in Paris. I am very happy to show my gratitude towards Dr. Giovanni Esposito for keeping the track of time during the course of my PhD and for his contant support on almost all administrative affairs. I would also like to thank him for his inputs in the activated sludge processes. Without his help, I am sure I will not be able to reach this stage. I also had the pleasure of attending very well organized summer school in Cassino, last summer with very good scientific discussions as well amazing food!

A very special thanks to Dr. Paul Mason and Dr. Erkan Sahinkaya, jury members, for their critical comments and questions on my thesis and defense. Your questions and comments have led me to think more deeply on my work and also provided me the perspective of my thesis. Moreover, it has also given me thoughts on how to improve this work further. I would also like to express my gratitude towards Prof. Stefan Uhlenbrook for chairing my PhD defense session.

I had the pleasue of working with Silvio, Dominic and Paolo during my thesis. I must thank them for all the fruitful discussions we had in "the office" and working enthusiastically on the topic. It has been a great pleasure in knowing you guys and thinking that I may have contributed something in making you great scientists! I would also like to thank my collaborators Dr. René Hubner, Dr. Harald Foerstendorf and Dr. Satoru Tsushima from HZDR, Germany for their expert comments during in many studies. I would also like to thank Stephan Weiss, HZDR, for carrying out many experiments and providing valuable inputs in the mansucripts in last 2 years. Many special thanks to Norbert Jordan with whom I collaborated for last 2 years on many studies. I would also like to thank him for being a great support and host during the course of this PhD and my stay in Dresden. It has been a pleasure knowing you and hoping to work with you for many coming years.

This journey would have been extremely hard without the support from colleagues and now friends Carlos, Lawrence, Javier, Niranjan, Angelica, Wook, Maribel, Mullelo, Mirjana, Jaka, and Salifu. It has been a lot of fun discussing science with Carlos, Lawrence and Javier in the laboratory. I would also like to thank Flavia, Wendy, Susma and Chiara for their support and encouragment in the laboratory as well as outside laboratory. This study has been a lot of fun with my selenium buddy, Erika. It had been great fun to share office and lab bench with you and, of course, many important "non scientific" discussions with you. Same goes to Lucian, whose love for selenium surpasses mine! Also, a very special thanks to Suthee who always have been my fall back guy, be it sampling, sample treatment or enjoying lunch at wook bar. I would also give my thanks to Marina for amazing discussion that we always had during summer schools and in Jardin des Plantes! it has been a pleasure knowing you and working with you. Also, a warm hearted thanks to Joana for introducing me to IHE in the beginning and our constant contact on facebook! It was also fun to meet Joy, who taught me how to be relax even in the middle of most trying circumstances. This section would be incomplete without thanking Pakshi and Venkata for introducing me to lab and getting over the rejected articles. And, many

many thanks to Eldon to be always there in need, you made my life easier as I could always count on you.

I would like to thank Ferdi and Berend for helping me in the lab but also introducing me to the world of very nice solvents! Thanks to Frank for his help in GF and the pumps. Without your help, this thesis would not have been completed. Thanks to you too Lyzette, for your support in the lab specially with the "always failing" IC . It has been a pleasure to know you Peter specially your one liners! Special thanks for that among many other things in the lab. A big thanks to Fred for his omnipresence to deal with every problem. Without you, I am not sure if I could have achieved half of what I have achieved. I would also like to thank Don for his support in the lab.

I would like to thank my parents, sister and family for supporting me to chase my dreams and stay away from them. Without your constant support, this would not have been possible. Also, Moi moi, your addition to our household has been an inspiring event and your relaxed attitude helped me stay relax during the trying times. This section will not be complete without thanking Purvi. You have been a great support for me during the PhD. Thanks for tolerating me during very long working hours in lab and writing hours in night. Thanks for living with me during my worst phases during the last three years. You have always been supportive and also to someone that I can look upto. Thanks for this. I learned patience and fortitude from you. It has been a pleasure to know and hope to keep knowing for the rest of my life.

Summary

Nanoparticles exhibit many unique properties as compared to bulk materials owning to their high surface to volume ratio. Elemental selenium nanoparticles also exhibit novel properties that are exploited in the fabrication of solar cells, semiconductor rectifiers and removal of mercury and copper. The chemical synthesis of elemental selenium nanoparticles is costly, requires specialized equipment and uses toxic chemicals. On the other hand, biological production of elemental selenium nanoparticles (BioSeNPs) can be a green replacement for their chemical synthesis.

BioSeNPs are produced by microbial reduction of selenite and selenate. The source of the selenium oxyanions can be wastewater, where microbial reduction is employed as a remediation technology for the removal of selenium. The formed BioSeNPs are colloidal poly-disperse particles with negative surface charge and are present in the effluent of the bioreactor. Thus far, the properties of these BioSeNPs are not very well understood. This knowledge will help to produce better quality selenium nanomaterials, exploit BioSeNPs applications in wastewater treatment and control the fate of these BioSeNPs in the microbial reactors and the environment.

The characterization of BioSeNPs revealed the presence of extracellular polymeric substances (EPS) on the surface of BioSeNPs. The EPS was identified to control the surface charge and to some extent the shape of the BioSeNPs. Microbial reduction at 55 and 65 °C can lead to the formation of selenium nanowires as compared to nanospheres when the reduction takes place at 30 °C. These selenium nanowires are present in a trigonal crystalline structure and form a colloidal suspension, unlike the chemically formed trigonal selenium nanorods. The colloidal nature is due to negative ζ-potential values owning to the presence of EPS on the surface of biogenic selenium nanowires. Since proteins are a major component present in the EPS, the presence of various proteins on the surface of BioSeNPs was determined. The interaction of various amino acids with the BioSeNPs was also evaluated.

The interaction of heavy metals and BioSeNPs was studied with a view of developing a technology where BioSeNPs present in the effluent of an upflow anaerobic sludge blanket (UASB) reactor are mixed with heavy metals containing wastewater leading to the removal of both BioSeNPs and heavy metals. It was found that Cu, Cd and Zn can be effectively adsorbed onto BioSeNPs. Cu was 4.7 times preferentially adsorbed onto BioSeNPs. The interaction of BioSeNPs with the heavy metals led to a less negative ζ-potential of BioSeNPs loaded with heavy metals and thus better settling of BioSeNPs.

The presence of BioSeNPs in the effluent of a bioreactor treating selenium oxyanions containing wastewaters is undesirable due to higher total selenium concentrations. When a UASB reactor was operated under mesophilic and thermophilic conditions, the total selenium concentration in the effluent under thermophilic conditions were lower than in the mesophilic bioreactor effluent, suggesting better trapping of BioSeNPs at elevated temperatures. When an activated sludge reactor system was investigated to aerobically reduce selenite to BioSeNPs and trap them in the activated sludge flocs, around 80% of the fed selenium was trapped in the biomass. Sequential extraction procedure revealed that the trapped selenium is in form of BioSeNPs. The trapping of BioSeNPs in the activated sludge improved its settleability and hydrophilicity.

Keywords: Selenium, bioreduction, BioSeNPs, EPS, ζ-potential, heavy metals, activated sludge, UASB reactors, thermophilic

Sommario

Le nanoparticelle presentano numerose e particolari caratteristiche, se paragonate ai materiali granulari, dovute alla loro elevata superficie specifica. Le particelle di selenio elementare, inoltre, presentano alcune nuove proprietà sfruttate nella produzione di celle solari e raddrizzatori semiconduttori di corrente e nella rimozione di mercurio e rame. La sintesi chimica di nanoparticelle di selenio elementare è costosa e richiede l'utilizzo di una strumentazione specifica e di reattivi tossici. D'altro canto, però, la produzione per via biologica di queste nanoparticelle (comunemente riconosciute con l'acronimo inglese "BioSeNPs") rappresenta un'alternativa "verde" per la loro sintesi chimica.

Le "BioSeNPs" sono prodotte in seguito alla riduzione dei seleniti e dei selenati ad opera di microrganismi. Un'ottima fonte di ossoanioni di selenio è costituita dalle acque reflue, all'interno delle quali l'attività dei microrganismi viene sfruttata come tecnologia per la rimozione del selenio. Le particelle così formate sono solidi colloidali con superficie caricata negativamente che si ritrovano negli effluenti dei bioreattori utilizzati. Finora le proprietà di queste particelle non sono ancora del tutto conosciute. Pertanto, una conoscenza più approfondita, consentirebbe di ottenere nanomateriali a base di selenio di qualità superiore, di sfruttare queste particelle al meglio per il trattamento delle acque reflue e, infine, di controllare il destino delle stesse particelle nei reattori biologici e nell'ambiente.

Nel presente studio, la caratterizzazione delle "BioSeNPs" ha rivelato la presenza di sostanze polimeriche extracellulari (EPS) sulla loro superficie. È stato dimostrato che gli EPS permettono di controllare la carica superficiale e, in qualche modo, anche la forma delle nanoparticelle. A 55 e 65°C, la cinetica biologica porta alla formazione di nanofili di selenio mentre, a 30°C, sono state ottenute nanosfere. I nanofili di selenio sono presenti con una struttura cristallina trigonale e formano una sospensione colloidale, a differenza dei "nanobastoncini" di forma trigonale che si formano per via chimica. La natura colloidale, infatti, è dovuta proprio alla presenza degli EPS che induce valori negativi di potenziale sulla superficie dei nanofili di selenio prodotti per via biologica. Numerose proteine sono state rivelate all'interno degli EPS. L'interazione tra differenti aminoacidi con le "BioSeNPs" è stata valutata.

L'interazione tra i metalli pesanti e le nanoparticelle di selenio prodotte biologicamente è stata studiata con lo scopo di sviluppare una tecnologia di rimozione simultanea di metalli, provenienti da un'acqua reflua, e "BioSeNPs", provenienti dall'effluente di un reattore del tipo UASB. Nel presente studio è stato osservato che rame, cadmio e zinco possono essere efficientemente adsorbiti sulla superficie delle "BioSeNPs", con il rame che ha presentato un'affinità con le nanoparticelle 4.7 volte superiore a quella degli altri due metalli studiati. L'interazione tra nanoparticelle di selenio e metalli pesanti ha fatto sì che si ottenesse una carica negativa inferiore sulla superficie delle particelle permettendo una migliore sedimentazione delle stesse.

La presenza delle "BioSeNPs" nell'influente di un bioreattore adibito al trattamento di acque contenenti ossoanioni del selenio non è desiderabile a causa delle elevate concentrazioni di selenio. Durante l'esercizio di un reattore UASB in condizioni mesofile e termofile, è stato osservato che la concentrazione di selenio totale nell'effluente era minore in condizioni termofile che in condizioni mesofile, suggerendo un miglior trattenimento delle nanoparticelle a temperature maggiori. Durante l'esercizio di un sistema a fanghi attivi finalizzato alla riduzione anaerobica dei seleniti in nanoparticelle e all'intrappolamento delle stesse nei fiocchi di fango, è stato osservato che circa l'80% del selenio alimentato al reattore è rimasto intrappolato nei fiocchi di fango attivo. Una procedura di estrazione sequenziale ha dimostrato che il selenio intrappolato era presente sotto forma di "BioSeNPs". La presenza di tali particelle nei fiocchi di fango attivo ha permesso di ottenere migliori caratteristiche di sedimentabilità e idrofilicità del fango stesso.

Parole chiave: Selenio, bio-riduzione, "BioSeNPs", EPS, ζ-potenziale, metalli pesanti, fango attivo, reattori UASB, termofilia

Résumé

Les nanoparticules présentent de nombreuses propriétés uniques par rapport aux matériaux grossiers et possèdent un rapport surface / volume élevée. Les nanoparticules de sélénium élémentaire présentent également de nouvelles propriétés qui sont exploitées dans la fabrication de cellules solaires, des redresseurs à semi-conducteurs et l'élimination du mercure et du cuivre. La synthèse chimique des nanoparticules de sélénium élémentaire est coûteuse, nécessite un équipement spécialisé et utilise des produits chimiques toxiques. D'autre part, la production biologique des nanoparticules de sélénium élémentaire (BioSeNPs) peut être une alternative verte pour leur synthèse chimique.

Les BioSeNPs sont produites par réduction microbienne de sélénite et séléniate. Les anions de sélénium peut être présent dans les eaux usées, où la réduction microbienne est utilisée comme une technique d'élimination du sélénium. Les BioSeNPs formés sont des particules colloïdales poly-dispersés avec une charge de surface négative et sont présents dans l'effluent du bioréacteur. Jusqu'à présent, les propriétés de ces BioSeNPs ne sont pas très bien comprises. Cette connaissance nous aidera à obtenir de meilleurs nanomatériaux de sélénium de qualité, exploiter des applications de BioSeNPs dans le traitement des eaux usées et contrôler le devenir de ces BioSeNPs dans les réacteurs microbiens et dans l'environnement.

La caractérisation de BioSeNPs a révélé la présence de substances polymériques extracellulaires (EPS) à la surface de BioSeNPs. Les EPS ont été identifiés comme étant capable de contrôler la charge de surface et dans une certaine mesure la forme des BioSeNPs. La réduction microbienne à 55 et 65°C peut conduire à la formation de nanofils de sélénium par rapport à nanosphères lorsque la réduction a lieu à 30 ° C. Ces nanofils de sélénium présentent une structure cristalline rhomboédrique et forment une suspension colloïdale, à la différence des nanofils de sélénium trigonal formés chimiquement. La nature colloïdale est due à des valeurs de potentiel ζ-négatives généré par la présence d'EPS à la surface de nanofils de sélénium biogènes. Étant donné que les protéines sont un composant majeur des EPS, la présence de différentes protéines sur la surface de BioSeNPs a été déterminée. L'interaction des différents acides aminés avec des BioSeNPs a également été évaluée.

L'interaction des métaux lourds avec les BioSeNPs a été étudiée en vue de l'élaboration d'une technologie où les BioSeNPs présents dans l'effluent d'un réacteur UASB est mélangés avec les métaux lourds contenus dans les eaux usées pouvant conduire à l'élimination simultanée des BioSeNPs et des métaux lourds. Il a été constaté que Cu, Cd et Zn peuvent être efficacement adsorbé sur BioSeNPs. Cu étant préférentiellement 4,7 fois plus adsorbé sur les BioSeNPs. L'interaction des BioSeNPs avec les métaux lourds conduit à un potentiel ζ moins négatif de BioSeNPs chargés de métaux lourds et ainsi une meilleure décantation des BioSeNPs.

La présence de BioSeNPs dans l'effluent du bioréacteur traitant des eaux usées contenant du sélénium est indésirable en raison des concentrations élevées de sélénium résiduel. Lorsqu'un réacteur UASB est utilisé dans des conditions mésophiles et thermophiles, la concentration résiduelle en sélénium dans l'effluent dans des conditions thermophiles ont été plus faibles que dans le bioréacteur effluent mésophile, ce qui suggère un meilleur piégeage de BioSeNPs à des températures élevées. Quand un réacteur à boues activées a été étudié pour réduire le sélénite en BioSeNPs en condition aérobie et piéger le BioSeNPs dans les flocs de boues activées, environ 80% du sélénium apporté a été piégé dans la biomasse. Une procédure d'extraction séquentielle a révélé que le sélénium est piégé sous forme de BioSeNPs. Le piégeage des BioSeNPs dans la boue activée améliore la décantation et le caractère hydrophile.

Mots-clés: sélénium, bio-réduction, BioSeNPs, EPS, potentiel ζ, métaux lourds, boues activées, réacteurs UASB, thermophiles

Samenvatting

Nanodeeltjes vertonen vele unieke eigenschappen in vergelijking met de eigenschappen van bulk materialen door hun hoge oppervlakte tot volume verhouding. Elementair selenium nanodeeltjes vertonen eveneens nieuwe eigenschappen die bij de vervaardiging van zonnecellen, halfgeleidergelijkrichters en de verwijdering van kwik en koper kunnen worden geëxploiteerd. De chemische synthese van elementair selenium nanodeeltjes is duur, vereist speciale infrastructuur en gebruikt giftige chemicaliën. Aan de andere kant kan de biologische productie van elementair selenium nanodeeltjes (BioSeNPs) een groen alternatief zijn voor hun chemische synthese.

BioSeNPs worden geproduceerd door microbiële reductie van seleniet en selenaat. De bron van de selenium anionen kan afvalwater zijn, waarbij microbiële reductie wordt toegepast als een saneringstechnologie voor de verwijdering van selenium. De gevormde BioSeNPs zijn colloïdaal poly-disperse deeltjes met een negatieve oppervlakteading en zijn aanwezig in het effluent van bioreactoren. Tot zover zijn de eigenschappen van deze BioSeNPs niet goed begrepen. Deze kennis kan bijdragen tot de productie van selenium nanomaterialen van betere kwaliteit, de exploitatie van BioSeNPs in afvalwaterzuiveringstoepassingen en het beïnvloeden van het lot van deze BioSeNPs in microbiële reactoren en het milieu.

De karakterisering van BioSeNPs toonde de aanwezigheid van extracellulaire polymere stoffen (EPS) op het oppervlak van de BioSeNPs aan. De EPS werden geïdentificeerd als regelaar van de oppervlakteading en gedeeltelijk ook van de vorm van de BioSeNPs. Microbiële reductie bij 55 en 65 °C kan leiden tot de vorming van selenium nanodraden, in tegenstelling tot de nanobolletjes gevormd wanneer de reductie plaatsvindt bij 30 °C. Deze selenium nanodraden zijn aanwezig in een trigonale kristalstructuur en vormen een colloïdale suspensie, in tegenstelling tot de chemisch gevormde trigonale selenium nanostaafjes. De colloïdale aard is te wijten aan de negatieve ζ-potentiaalwaarden van het oppervlak, veroorzaakt door de aanwezigheid van EPS op het oppervlak van de biogene selenium nanodraden. Aangezien eiwitten een belangrijke component in de EPS zijn, werd de aanwezigheid van verschillende eiwitten op het oppervlak van de BioSeNPs bepaald. De interactie van verschillende aminozuren met de BioSeNPs werd ook geëvalueerd.

De interactie van zware metalen en BioSeNPs werd bestudeerd teneinde een technologie te ontwikkelen waarin BioSeNPs aanwezig in het effluent van een opwaartse anaërobe slibdeken (UASB) reactor gemengd worden met zware metalen bevattend afvalwater, wat leidt tot de verwijdering van zowel BioSeNPs en zware metalen. Er werd gevonden dat Cu, Cd en Zn effectief kunnen worden geadsorbeerd door BioSeNPs. Cu was 4,7 keer preferentieel geadsorbeerd door de BioSeNPs. De interactie van de BioSeNPs met de zware metalen leidt tot een minder negatieve ζ-potentiaal van de met zware metalen beladen BioSeNPs en dus betere bezinking van deze BioSeNPs.

De aanwezigheid van BioSeNPs is ongewenst in het effluent van een bioreactor die selenium oxyanionen bevattend afvalwater behandeld vanwege de hogere totale selenium concentraties in het effluent. Wanneer UASB reactoren onder mesofiele en thermofiele omstandigheden werden bedreven waren de totale selenium concentraties in het effluent onder thermofiele omstandigheden lager dan in het mesofiele bioreactoreffluent. Dit suggereert dat de BioSeNPs bij verhoogde temperaturen beter worden ingevangen. Wanneer de capaciteit van een actief slib reactor werd onderzocht om seleniet aëroob te verwijderen, bleek dat de BioSeNPs in de actief slibvlokken werden ingevangen en ongeveer 80% van het toegevoerde selenium werd in de biomassa weerhouden. Sequentiële extractieprocedures gaven aan dat het ingevangen selenium aanwezig was in de vorm van BioSeNPs. Het invangen van BioSeNPs in het actief slib verbeterde de bezinkbaarheid en de hydrofiliciteit van de aktief slibvlokken.

Keywords: Selenium, bioreductie, BioSeNPs, EPS, ζ-potentiaal, zware metalen, actief slib, UASB reactoren, thermofiel

CHAPTER 1

Introduction

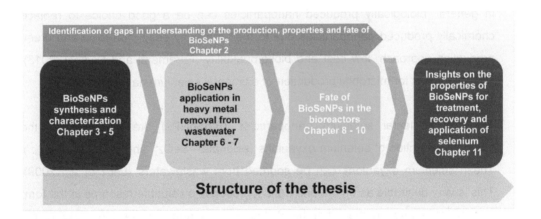

1.1 Background

Elemental selenium displays many unique properties such as high photoconductivity, piezoelectric, thermoelectric and non-linear electric response (Gates et al., 2002). These properties are enhanced when the elemental selenium is in the form of nanoparticles due to high surface to volume ratio. Indeed, selenium nanomaterials have been used in xerography, solar cells, semiconductor rectifiers and other functional materials. Elemental selenium nanowires are excellent systems for studying size confinement effect on optical, electrical and mechanical properties of elemental selenium or use them as connectors in fabrication of nanodevices. Elemental selenium nanospheres have been used to capture mercury and copper from vapor and aqueous phase, respectively (Bai et al., 2011; Johnson et al., 2008).

Elemental selenium nanoparticles can be fabricated using methods such as physical vapour deposition, vapour phase diffusion and wet chemical methods (Chen et al., 2010; Ma et al., 2008; Shah et al., 2010; Stroyuk et al., 2008). However, these methods are usually costly, use toxic solvents and produce hazardous by-products. In general, biologically produced nanoparticles can be a good choice to replace chemically produced nanoparticles due to their relatively easy, non-toxic and green production approach, low cost and biocompatibility (Faramarzi and Sadighi, 2013). This also holds for microbial production of elemental selenium nanoparticles.

Biogenic elemental selenium nanoparticles (BioSeNPs) are produced by the microbial reduction of selenium oxyanions (selenate - SeO_4^{2-} and selenite - SeO_3^{2-}). These selenium oxyanions can be sourced from wastewaters (Lenz et al., 2009). This is very desirable as on the one hand it produces a valuable resource in the form of BioSeNPs and on the other hand this process removes selenium oxyanions, which are toxic and bioavailable, from wastewaters (Lenz et al., 2008). The toxicity of selenium depends largely on its speciation. The dissolved form of selenium, selenate and selenite are known to be toxic due to their higher bioavailability. On the other hand, selenides (oxidation states -2, -1) are unstable and rapidly react to form metal selenide or oxidize to form elemental selenium (Lenz and Lens, 2009). The elemental form of the selenium is known to be stable over a large range of pH and redox conditions as well as to be less bioavailable as compared to selenate or selenite (Winkel et al., 2012).

1.2 Problem description

The microbial production has always led to the formation of colloidal, polydisperse spherical BioSeNPs with average diameter exceeding 100 nm, thus impacting their applications as a nanomaterial (Oremland et al., 2004). There is not enough understanding on the factors governing the size and shape of BioSeNPs. Also, no study has so far been carried out on the production of biogenic selenium nanowires and to characterize these.

From an environmental perspective, BioSeNPs are a "double-edged sword". On the one hand, BioSeNPs can be used to adsorb mercury from the vapor phase and on the other hand their presence in the effluent of the bioreactors increases the total selenium concentration leading to the requirement of a second additional step to treat selenium oxyanions containing wastewaters (Buchs et al., 2013). However, there are no studies that have explored the full potential of BioSeNPs in the heavy metal removal nor there are studies that have attempted to understand the fate of BioSeNPs in bioreactors.

Selenium is a scarce resource with its application in industry and is required in human diet (Haug et al., 2007). The elemental form of selenium is much more widespread than previously thought and can constitute 30-60% of the total selenium content in the sediments (Zhang et al., 2004). Microbial transformation of selenium oxyanions to BioSeNPs is widespread and contributes greatly to the global cycle of selenium (Vriens et al., 2014; Winkel et al., 2012). However, our understanding on the factors governing the properties of BioSeNPs is very limited. Our improved understanding can help us to better predict, recover and reuse selenium.

1.3 Research objective

The main objective of this thesis is to develop a better understanding of the production process and properties of BioSeNPs and to explore their potential and relevance in wastewater treatment.

The specific objectives are:

1) To produce and characterize BioSeNPs produced by anaerobic granular sludge:
    a) To identify the origin and effect of the organic layer present on the surface of BioSeNPs on their properties
    b) To produce and characterize biogenic elemental selenium nanowires (BioSeNWs)
    c) To study the interaction of amino acids with elemental selenium in the BioSeNPs

2) To explore the potential of BioSeNPs in the removal of heavy metals from wastewaters:
    a) To assess BioSeNPs capability to adsorb the model heavy metal Zn and the effect of adsorption of Zn onto BioSeNPs on their colloidal stability
    b) To assess BioSeNPs capability to selectively adsorb heavy metals from an equimolar mixture of Cu, Zn and Cd

3) To study the fate of BioSeNPs in aerobic and anaerobic wastewater treatment bioreactors:
    a) To assess the thermophilic conditions in an upflow anaerobic sludge blanket reactor (UASB) for better retention of BioSeNPs in the reactor
    b) To optimize the operating conditions in the activated sludge reactor to trap maximum BioSeNPs in the biomass
    c) To characterize the activated sludge fed with selenium for their physicochemical properties

1.4 Structure of the thesis

This dissertation comprises eleven chapters. The thesis is divided into three main sections: BioSeNPs synthesis and characterization (Chapters 3 - 5), BioSeNPs application in heavy metal removal from wastewater (Chapters 6 and 7) and fate of BioSeNPs in the bioreactors (Chapter 8 - 10) (Figure 1.1).

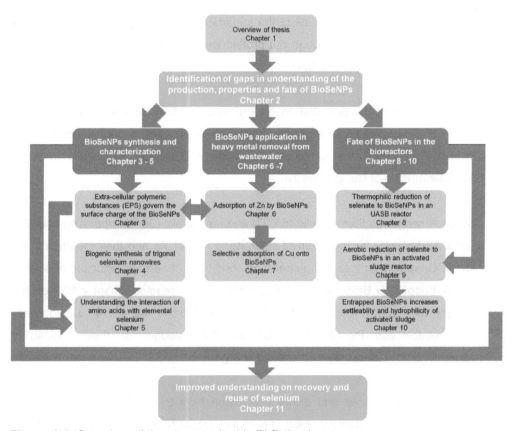

Figure 1.1. Overview of the chapters in this PhD thesis

The following paragraph provides details about the composition of this thesis

Chapter 1 provides a general overview of this dissertation that includes background, problem statement, research objectives and thesis structure.

Chapter 2 describes the state-of-art of BioSeNPs production methods and their characterization. This chapter also discusses the challenges in the production of BioSeNPs.

Chapter 3 explores the origin of the organics present on the surface of BioSeNPs. This chapter also identifies the effect of the organic layer capping the BioSeNPs on their surface charge, shape and size. Chapter 4 presents the production and characterization of BioSeNWs by use of thermophilic conditions. Chapter 5 explores the relative affinity of amino acids to the surface of elemental selenium by means of statistical analysis of high-through put protein identified by Hybrid Ion Trap-Orbitrap Mass Spectrometer (LTQ-Orbitrap).

Chapter 6 explores the potential adsorbent capacity of BioSeNPs towards Zn. This chapter also observed the effect on ζ-potential of BioSeNPs upon loading of Zn. Chapter 7 describes the selective adsorption of Cu onto BioSeNPs from the equimolar mixture of Cu, Cd and Zn.

Chapter 8 compares thermophilic (55 $^{\circ}$C) and mesophilic (30 $^{\circ}$C) operating condition in an UASB reactor for reduction of selenate to BioSeNPs and their retention in the reactor. Chapter 9 optimizes the operating conditions of aerobic reduction of selenite and the trapping of BioSeNPs in the activated sludge reactor. Chapter 10 identifies the speciation and properties of the trapped selenium in the activated sludge and also characterizes the selenium fed activated sludge for their settleability, hydrophilicity, dewaterability and surface charge.

Chapter 11 summarizes and draws conclusion from this study. The chapter also provides future recommendations and perspectives for further research.

1.5 References

Bai, Y., Rong, F., Wang, H., Zhou, Y., Xie, X., Teng, J., 2011. Removal of copper from aqueous solution by adsorption on elemental selenium nanoparticles. J. Chem. Eng. Data 56, 2563–2568.

Buchs, B., Evangelou, M.W.-H., Winkel, L., Lenz, M., 2013. Colloidal properties of nanoparticular biogenic selenium govern environmental fate and bioremediation effectiveness. Environ. Sci. Technol. 47, 2401–2407.

Chen, H., Shin, D., Nam, J., Kwon, K., Yoo, J., 2010. Selenium nanowires and nanotubes synthesized via a facile template-free solution method. Mater. Res. Bull. 45, 699–704.

Faramarzi, M.A., Sadighi, A., 2013. Insights into biogenic and chemical production of inorganic nanomaterials and nanostructures. Adv. Colloid Interface Sci. 189-190, 1–20.

Gates, B.B., Mayers, B., Cattle, B., Xia, Y., 2002. Synthesis and characterization of uniform nanowires of trigonal selenium. Adv. Funct. Mater. 12, 219–227.

Haug, A., Graham, R.D., Christophersen, O.A., Lyons, G.H., 2007. How to use the world's scarce selenium resources efficiently to increase the selenium concentration in food. Microb. Ecol. Health Dis. 19, 209–228.

Johnson, N.C., Manchester, S., Sarin, L., Gao, Y., Kulaots, I., Hurt, R.H., 2008. Mercury vapor release from broken compact fluorescent lamps and in situ capture by new nanomaterial sorbents. Environ. Sci. Technol. 42, 5772–5778.

Lenz, M., van Aelst, A.C. Van, Smit, M., Corvini, P.F.X., Lens, P.N.L., 2009. Biological production of selenium nanoparticles from waste waters. Mater. Res. 73, 721–724.

Lenz, M., Hullebusch, E.D.V, Hommes, G., Corvini, P.F.X., Lens, P.N.L., 2008. Selenate removal in methanogenic and sulfate-reducing upflow anaerobic sludge bed reactors. Water Res. 42, 2184–2194.

Lenz, M., Lens, P.N.L., 2009. The essential toxin: the changing perception of selenium in environmental sciences. Sci. Total Environ. 407, 3620–33.

Ma, J., Liu, X., Wu, Y., Peng, P., Zheng, W., 2008. Controlled synthesis of selenium of different morphologies at room temperature. Cryst. Res. Technol. 43, 1052–1056.

Oremland, R.S., Herbel, M.J., Blum, J.S., Langley, S., Beveridge, T.J., Ajayan, P.M., Sutto, T., Ellis, A. V, Curran, S., 2004. Structural and spectral features of selenium nanospheres produced by se-respiring bacteria. Appl. Environ. Microbiol. 70, 52–60.

Shah, C.P., Dwivedi, C., Singh, K.K., Kumar, M., Bajaj, P.N., 2010. Riley oxidation: A forgotten name reaction for synthesis of selenium nanoparticles. Mater. Res. Bull. 45, 1213–1217.

Stroyuk, A.L., Raevskaya, A.E., Kuchmiy, S.Y., Dzhagan, V.M., Zahn, D.R.T., Schulze, S., 2008. Structural and optical characterization of colloidal Se nanoparticles prepared via the acidic decomposition of sodium selenosulfate 320, 169–174.

Vriens, B., Lenz, M., Charlet, L., Berg, M., Winkel, L.H.E., 2014. Natural wetland emissions of methylated trace elements. Nat. Commun. 5, 3035.

Winkel, L.H.E., Johnson, C.A., Lenz, M., Grundl, T., Leupin, O.X., Amini, M., Charlet, L., 2012. Environmental selenium research: from microscopic processes to global understanding. Environ. Sci. Technol. 46, 571–579.

Zhang, Y., Zahir, Z.A., Frankenberger Jr., W.T. Jr., 2004. Fate of colloidal-particulate elemental selenium in aquatic systems. J Env. Qual 33, 559–564.

CHAPTER 2
Literature review

This chapter has been published as

Jain, R.; Gonzalez-Gil, G.; Singh, V., van Hullebusch, E.D., Farges, F.; Lens, P.N.L., 2014. Biogenic selenium nanoparticles, Production, characterization and challenges. In Kumar, A., Govil, J.N., Eds. Nanobiotechnology. Studium Press LLC, USA, pp. 361-390.

Abstract:

Selenium nanoparticles can be readily produced by microbial reduction of selenium oxyanions under anaerobic as well as aerobic conditions. This method is advantageous as the product can be produced at ambient temperature and pressure with relatively non specialized equipment. Moreover, the biogenic selenium nanoparticles demonstrate unique optical and spectral properties. However, the biogenic selenium nanoparticles are polydisperse and their size (> 30 nm) is on the larger side for applications. Also, in many cases, the biogenic selenium nanoparticles have to be separated from the biomass, leading to increased production time and costs. Synthetic biology can help us to better understand the mechanism and pathway of selenium nanoparticles production and eventually help us to improve or design micro-organisms those can produce selenium nanoparticles with desired properties.

Key words: Biogenic, selenium, nanoparticles, characterization, proteins, synthetic biology

Graphical abstract:

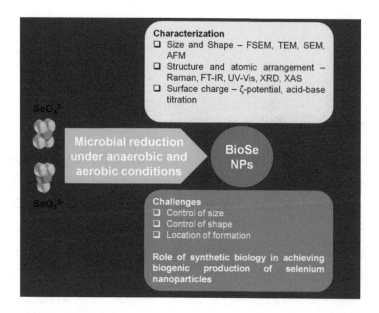

2.1. Introduction

Nanotechnology is the science of developing and utilizing materials, systems or devices at roughly 1 - 100 nm scale. According to European Union official definition, "50% or more of the particles in the number size distribution, one or more external dimensions is in the size range 1 nm-100 nm" can be termed as nanomaterial (EU recommendations, 2011). At these scales, materials, devices and systems exhibit novel optical, electrical, photo-electrical, magnetic, mechanical, chemical and biological properties those are different from their bulk properties. The essence of nanotechnology is to use these nano-blocks to build larger structures which are fundamentally new materials with unique properties (Walsh et al., 2008; Qu et al., 2013). Nanoscale materials have various applications in electronics, sensing devices, drug delivery, medicine and photonics.

Due to the unique properties of selenium nanoparticles, there is an interest in their production for nanotechnology applications. For example, research is being carried out to use selenium nanoparticles for medicinal purposes such as antifungal applications, anti-cancer orthopedic implants or treatment of malignant mesothelioma (Webster, 2007; Shahverdi et al., 2010). Nanowires formed by selenium nanoparticles demonstrate novel photoconductivity (Gates et al., 2002) and amorphous selenium nanoparticles have shown unique photoelectric, semiconducting and X-ray-sensing properties (Smith and Cheatham, 1980). These nanomaterials can be exploited in nanowire electronics, sensors and more efficient solar cells. From an environmental perspective, selenium nanoparticles have been shown to capture mercury from the gaseous phase and precipitate on nanoparticles' surface as HgSe (Johnson et al., 2008; Fellowes et al., 2011).

2.2. Production of selenium nanoparticles

Selenium nanoparticles can be produced using the biological or chemical methods. Chemical production methods include reduction of sodium selenite by glutathione (GSH, glutamylcysteinylglycine) (Johnson et al., 2008) or glucose (Chen et al., 2010), by reaction of ionic liquid with sodium selenosulfate (Langi et al., 2010) and various other approaches (Abdelouas et al., 2000; Gates et al., 2002; Ma et al., 2008; Shah et al., 2010; Shah et al., 2010a; Zhang et al., 2010; Dwivedi et al., 2011).

Chemical methods produce selenium nanoparticles of desired size and polydispersity index as reported in several studies (Johnson et al., 2008; Langi et al., 2010). However, these methods are expensive, environmentally hazardous and in many cases require specialized equipment. On the other hand, the biological production methods are simple and can be carried out at ambient temperature and pressure (Oremland et al., 2004). There are numerous species of archaea and bacteria present in nature those can reduce selenate or selenite to produce colloidal elemental selenium (Oremland et al., 2004; Stolz et al., 2006).

A study by Oremland et al., (2004) compares some features of biologically and chemically produced selenium nanoparticles. The authors show that monoclinic crystalline structures of selenium nanoparticles produced by selenium oxyanion respiring bacteria were compact, uniform, stable and their size ranged from 200 to 400 nm. In contrast, the size of selenium nanoparticles produced by auto oxidation of H_2Se gas and chemical reduction of selenite with ascorbate ranged between 10 nm to 50 □m. Moreover, all the three different microbial species - *Sulfurospirillum barnesii*, *Bacillus selenitireducens* and *Selenihalanaerobacter shriftii* used in this study, showed unique and different optical properties. The band gap energy, the energy required to excite a valence electron to the conduction electron, was lower for all three biologically synthesized nanospheres compared to chemically synthesized nanospheres. The low band gap energy gives a promising option for biologically synthesized nanoparticles to be used in solar cells, rectifier and xerography. This finding opens doors of opportunities to synthesize selenium nanoparticles biologically with unique structural and optical properties.

2.2.1. Biological Production of Selenium Nanoparticles

There are many species of bacteria, archaea and plants those produce selenium nanoparticles by reducing selenium oxyanions, *i.e.* selenate - SeO_4^{2-} and selenite - SeO_3^{2-} (Lenz and Lens, 2009). Various bacteria and archaea have been reported to couple their growth to the reduction of selenite/selenate (*i.e.* dissimilatory reduction).

Under anaerobic conditions, dissimilatory reduction is the main metabolic process for production of selenium nanoparticles (Oremland et al., 2004; Stolz et al., 2006). Under aerobic conditions, redox poise (Yamada et al., 1997) and detoxification (Lortie et al., 1992; Dhanjal and Cameotra, 2010) are the main mechanisms. Fungi also reduce selenium oxyanions to elemental selenium nanoparticles as a method of detoxification. However, other than reduction, fungi can also take up and/or biomethylate selenium oxyanions to volatile derivatives of selenium, though these methods do not produce selenium nanoparticles (Gharieb, 1995).

2.2.1.1. Biogenic production of selenium nanoparticles under aerobic conditions

Kuroda et al., 2011) used *Pseudomonas stutzeri* to explore the effect of temperature, pH and NaCl concentration on selenate and selenite reduction rates (Figure 2.1) under aerobic conditions (Table 2.1. Yadav et al., 2008) reported the formation of amorphous elemental selenium under aerobic conditions by the soil bacterium *Pseudomonas aeruginosa*. The growth rate of *Pseudomonas aeruginosa* in the presence of 5, 15 and 25 mg/L (0.029, 0.087 and 0.145 mM Se) of sodium selenite was comparable to the growth rate without sodium selenite in the medium.

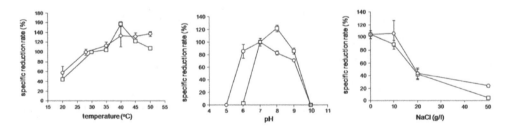

Figure 2.1. Effect of temperature, pH and salinity on specific selenate and selenite reduction rates by *Pseudomonas stutzeri* (open square - selenate; open circles – selenite) (Reproduced with permission from Kuroda et al., 2011).

Table 2.1. Micro-organisms capable of producing selenium nanoparticles by reduction of selenite or selenate under aerobic or anaerobic conditions. The incubation conditions, the maximum concentration of selenite or selenate and the size of the produced selenium nanoparticles are compiled.

Micro-organism	Conditions	SeO_4^{2-} (mM)	SeO_3^{2-} (mM)	End product and Size of selenium particles (d. nm)	References
Pseudomonas stutzeri	Aerobic; see Fig. 2.1	48 & 122	19 & 94	Se(0), <200	Lortie et al. (1992); Kuroda et al. (2011)
Pseudomonas aeruginosa	Aerobic; T = 28 °C; pH 5.5 - 6	NR	0.144	Se(0), NR	Yadav et al. (2008)
Pesudomonas alcaliphila	Aerobic; T = 28 °C; pH 7.5	NR	100	Se(0), 50 – 500	Zhang et al. (2011)
Pseudomonas fluorescens	Aerobic; T = 26 °C; pH neutral	NR	0.2	Se(0), NR	Belzile et al. (2006)
Bacillus sp.	Aerobic; Room temperature	NR	1	Se(0), 100 – 200	Tejo Prakash et al. (2009)
Bacillus cereus	Aerobic; T = 37 °C	NR	10	Se(0), 150 - 200	Dhanjal and Cameotra. (2010)
Bacillus megaterium	Aerobic; T = 37 °C; pH 7.5	NR	2	Se(0), ~ 200	Mishra et al. (2011)
Bacillus subtilis	Aerobic; T = 35 °C; pH 7.0	NR	4	Se(0), NR	Wang et al. (2010)
Rhizobium sp.	Aerobic; T = 28 °C NO_3^{1-} = 10 mM	NR	5	Se(0), NR	Hunter et al. (2007)
Enterobacter cloacae	Open to atm.; T = 28 °C	0.6		Se(0), NR	Losi and Frankenberger. (1997)
Bacillus selenitireducens	Anaerobic; T = 25 °C; pH 9.8; Salinity 56 g/L	NR	3 & 10	Se(0), 200 - 400	Oremland et al., (2004)
Sulfurospirillu	Anaerobic; T =	20	3	Se(0), 200 -	Oremland et al.

m barnesii	25 °C, pH 7.3; Salinity 2g/L			400	(2004)
Selenihalanaer obacter shriftii	Anaerobic; T = 25 °C; pH 7.0; Salinity 205g/L	20	3	Se(0), 200 - 400	Oremland et al. (2004)
Shewanella sp. HN-41	Anaerobic; T = 30 °C	NR	0.1 & 0.5	Se(0), Fig. 2.4	Lee et al. (2007); Tam et al. (2010)
Shewanella oneidensis MR-1	Aerobic, anaerobic	NR	1 & 2	Se(0), NR	Klonowska et al. (2005)
Anaerobic granules	Anaerobic; T = 30 °C; pH 7.0; HRT = 6 hours, Superficial velocity = 1m/hr	10 & 0.04 mM/day		Se(0), 150; 100 - 500	Lenz et al. (2008a); Lenz et al. (2008b)
Klebsiella pneumoniae	Anaerobic; T = 37 °C pH 7.2	NR	3.7	Se(0), 100 - 500	Fesharaki et al. (2010)
Rhodospirillum rubrum	Aerobic; anaerobic; T = 30 °C, incandescent light (35 W/m^2)	NR	0.5 & 2	Se(0), NR	Kessi et al. (1999)
Azospira oryzae	Micro-aerophillic; anaerobic; T = 28 °C	10	4	Se(0), NR	Hunter. (2007)
Veillonella atypica	Anaerobic; T = 37 °C; pH 7.5	NR	5	Se(0) - 120, ZnSe - 30	Pearce et al.(2008)

NR - Not reported

15

Bacillus cereus, isolated from coalmine soils and later identified on the basis of morphological, biochemical and molecular methods, produced selenium nanoparticles by reduction of selenite (Dhanjal and Cameotra, 2010). The microorganism was grown between 0.5 mM to 10 mM of sodium selenite and its growth profile was found to be comparable to that of *Bacillus cereus* when grown without selenite stress. However, the size of the bacteria after 48 h of growth in selenite containing medium was smaller than the size of the bacteria grown without selenite stress (Figure 2.2). *Bacillus* species, that showed the 99% 16S rRNA gene sequence homology to *Bacillus thuringiensis, B. anthracis,* and *B. cereus,* produced selenium nanoparticles only under the aerobic conditions (Tejo Prakash et al., 2009).

A strain belonging to the genus *Rhizobium,* with its 16S rRNA sequence more than 2.7% different than that of *R. radiobacter* or *R. rubi,* was able to reduce selenite to elemental selenium under the aerobic conditions (Hunter et al., 2007). The rate of selenite reduction improved when nitrate was present with selenite. During the reduction of selenite in the presence of nitrate, reduction of nitrate and accumulation of nitrite was also observed. However, this microorganism was unable to reduce selenate, either in the presence or in the absence of nitrate.

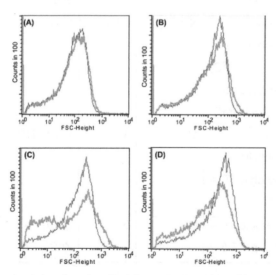

Figure 2.2. Flow cytometry showing that the average size of bacterial cells gradually decreases when *Bacillus cereus* was grown in the presence of selenite oxyanions.

The size of bacterial cell is smaller when there is a shift towards the left side on the log-scaled-X axis and *vice versa*. T indicates test population (with selenite) and C indicates control population without selenite., (A) After 12 h of incubation., (B) After 24 h of incubation., (C) After 36 h of incubation and (D) After 48 h, the cell size of test population decreased as compared to the control population evidencing the stress caused by selenite (Reproduced with permission from Dhanjal and Cameotra, 2010).

2.2.1.2. Biogenic production of selenium nanoparticles under anaerobic conditions

Bacillus selenitireducens was isolated for the first time from Mono Lake, California, USA (Switzer et al., 1998). It was able to reduce 10 mM of sodium selenite in less than 80 h and produced equivalent amounts of spherical selenium nanoparticles (Oremland et al., 2004). Apart from selenite, *Bacillus selenitireducens* was capable to utilizing 5 - 10% oxygen, arsenate, fumurate, thiosulfate, trimethylamine oxide, nitrate and nitrite as electron acceptor. And besides lactate, it can utilize pyruvate, starch, fructose, galactose and glucose as electron donor.

Shewanella species HN-41 was used to study the effect of the initial biomass concentration, reaction time and initial selenite concentration on the selenium nanoparticles size and formation rate (Tam et al., 2010). As expected, the higher initial biomass concentrations led to higher selenite reduction rates. This study also evaluated the effect of the initial selenite concentration on the production of selenium nanoparticles or reduction of selenite. A concentration of 0.1 mM selenite was required to achieve the maximum selenite reduction rates under the conditions tested and increased initial selenite concentrations did not lead to higher selenite reduction rates or higher selenium nanoparticles production rates. The kinetics of the reaction was well described by a Michaelis-Menten relationship with estimated values for the V_{max} and K_m of 1.37μM/h and 88 μM, respectively.

Shewanella oneidensis has been studied for the effect of various electron donors and the presence of other anions on the selenite reduction rates (Klonowska et al., 2005). The highest selenite reduction rate was obtained when this bacterium was grown with Luria-Bertani or in the presence of yeast extract. The second highest reduction rate was obtained with lactate (13%) of selenite reduction rate obtained when using Luria-Bertani as medium. The presence of other anions such as nitrate, nitrite, fumarate, TMAO (trimethylamine-N-oxide), and dimethyl sulfoxide resulted in almost 95% inhibition of the selenite reduction rate.

Different types of anaerobic granular sludge, suspended sludge, soils and sediments were studied for their potential to remove selenate from wastewater (Astratinei et al., 2006) 400 to 1500 μg gVSS^{-1} h^{-1} of selenate removal was achieved. In a study by Lenz et al., (2008a), anaerobic granules obtained from a full scale Upflow Anaerobic Sludge Bioreactor (UASB) reactor was used to reduce selenate oxyanion and selenium nanoparticles of 100 - 500 nm in diameter were obtained. In similar study by Lenz et al., (2008b), selenium nanoparticles of 50 - 100 nm were found on biofilms growing in the tubes of the bioreactor. Furthermore, precipitates of elemental selenium (approximately 150 nm in size) were found between the bioreactor and the settler tubes. The selenate removal efficiency in this study remained at values exceeding 91.9% and the COD removal efficiencies remained stable at 85% when there was no reactor disturbance such as lowering of the operating temperature or a decrease in the superficial upflow velocity.

A facultative anaerobic bacterium, *Klebsiella pneumoniae* reduced selenite to produce selenium nanoparticles (Fesharaki et al., 2010). Among various broths tested, the highest selenite reduction capacity of *Klebsiella pneumoniae* was observed in Tryptic Soy broth (TSP) at 1.92 mg Se/mL followed by Muller-Hinton broth (1.12 mg Se/mL), Luria-Bertani broth (0.96 mg Se/mL) and Nutrient broth (0.26 mg Se/mL).

Other type of selenium nanoparticles such as zinc selenide (ZnSe) and cadmium selenide (CdSe) nanoparticles are known for their non-linear optics, luminescence, electronics and catalyst properties. *Veillonella atypica* has been shown to produce ZnSe and CdSe nanoparticles (Pearce et al., 2008). In this study, 5 mM of sodium selenite was added as electron acceptor and 75 mM of sodium acetate, sodium lactate, sodium formate or hydrogen (in the head space) as electron donor. A reduction rate of 252 mM g^{-1} (biomass) h^{-1} was obtained for hydrogen as sole electron donor in the presence of 100 μM anthraquinone-2,6-disulfonate (AQDS).

The study also demonstrated that the presence of 100 µM of the soluble redox mediator AQDS improved the selenite reduction rate from 36 mM g^{-1} (biomass) h^{-1} to 252 mM g^{-1} (biomass) h^{-1} using hydrogen as sole electron donor. The selenite reduction rate was the lowest with sodium lactate (0.03 mM g^{-1} biomass h^{-1}) as electron donor. In this study, selenite was reduced to elemental selenium and then further reduced to produce selenide. The selenium nanoparticles, produced as an intermediate, were of approximately 120 nm in size. Once $ZnCl_2$ was added to the medium, ZnSe particles were formed. The ZnSe nanoparticles were 27 nm in size.

2.3. Characterization of selenium nanoparticles

Relevant characteristics of selenium nanoparticles that determine their applicability in nanotechnology are composition, size, shape, structure, atomic arrangement and surface charge. The composition of selenium nanoparticles is mainly characterized by using energy dispersive X-ray spectroscopy and X-ray photoelectron spectroscopy (Oremland et al., 2004; Tejo Prakash et al., 2009; Wang et al., 2010). The size and shape of selenium nanoparticles are determined by field emission scanning electron microscopy, transmission electron microscopy and atomic force microscopy (Oremland et al., 2004; Tejo Prakash et al., 2009; Wang et al., 2010; Dhanjal and Cameotra, 2010). The structure of selenium nanoparticles is determined by a combination of techniques such as Raman spectroscopy, Fourier transform infra red spectroscopy, UV-visible spectroscopy and X-ray diffraction (Oremland et al., 2004; Wang et al., 2010). More detailed analysis such as the atomic arrangement of selenium nanoparticles can be determined by X-ray absorption spectroscopy (van Hullenbusch et al., 2007; Lee et al., 2007; Pearce et al., 2008; Lenz, 2008; Lenz et al., 2008c; Lenz et al., 2011a).

2.3.1. Elemental composition of selenium nanoparticles

In all studies, the biologically produced selenium nanoparticles were entirely composed of elemental selenium (Figure 2.3; Oremland et al., 2004; Dhanjal and Cameotra, 2010; Fellowes et al., 2011) with an exception (Pearce et al., 2008) in which metal selenide nanoparticles were produced.

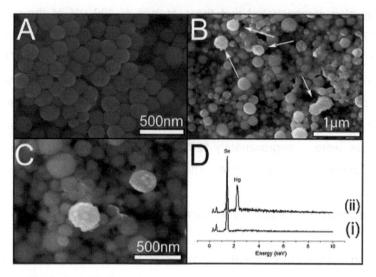

Figure 2.3. SEM and EDX spectra of selenium nanoparticles and selenium nanoparticles exposed to Hg a) Se nanoparticles produced by *G. sulfurreducens*, b) HgSe precipitation on surface of Se nanoparticles, c) Higher magnification of image b, d) EDX spectra of Se nanoparticles (i) and Se nanoparticles after being exposed to Hg vapor (ii) (Reproduced with permission from Fellowes et al., 2011).

2.3.2. Size and shape of selenium nanoparticles

Biogenic selenium nanoparticles, produced by all the reported micro-organisms, were spherical in shape and in some cases transformed from spherical particles to nanowires. However, the size of the biogenic nanoparticles differs depending on the production time and the type of micro-organism reducing selenium oxyanions. All microorganisms studied so far produce polydisperse nanoparticles with size ranging from 50 nm to 500 nm. The average size is always above 100 nm (Oremland et al., 2004; Tejo Prakash et al., 2009; Dhanjal and Cameotra, 2010; Kuroda et al., 2011; Zhang et al., 2011).

The effect of the temperature and oxygen concentration on the shape and size of selenium nanospheres produced by *Shewanella* sp HN-41 was studied (Lee et al., 2007). The average size of the nanospheres was higher at elevated (4, 15 and 30 $^{\circ}$C) temperatures. Conversely, when the O_2 concentration in the medium was increased, the average size of the selenium nanospheres decreased (Figure 2.4) and the shape of the particles became more irregular. These results suggest that production of size-controlled biological selenium nanospheres may be achieved by simply changing the culture conditions. The effect of the initial biomass concentration, reaction time and initial selenite concentration was systematically investigated on the size distribution and formation rate of selenium nanoparticles produced by *Shewanella sp* HN-41 (Tam et al., 2010). The initial biomass concentration did not affect the average size of the particles but affected their size distribution to a small extent. Over time, the average size of the selenium nanoparticles increased from 35 - 40 nm (2 h) to 120 nm (12 h). The initial selenite concentration (0.1 mM to 1.0 mM) had no effect on the particles size. In another study using *Pseudomonas alcaliphila*, selenium nanoparticles increased in size from 50 - 200 nm at 12 h of reaction time to 500 - 600 nm selenium particles after 24 h of reaction, indicating that the particles grew via Ostwald ripening process (Zhang et al., 2011).

It has been reported that *Veillonella atypica* reduced selenite to selenium nanospheres of 120 nm (Pearce et al., 2008). The reduction process continued and led to the formation of selenide in the system and when $ZnCl_2$ was added, it lead to the formation of ZnSe particles of 27 nm in size. However, $ZnCl_2$ particles of 27 nm are too large for quantum dot applications. To further decrease the size of ZnSe, biogenic selenide was extracted and a simple wet chemical reaction with Zn and Cd was carried out in the presence of thiol as a capping agent. Size variation between 3-6 nm was achieved for ZnSe particles and 2-4 nm for CdSe nanoparticles (Pearce et al., 2008).

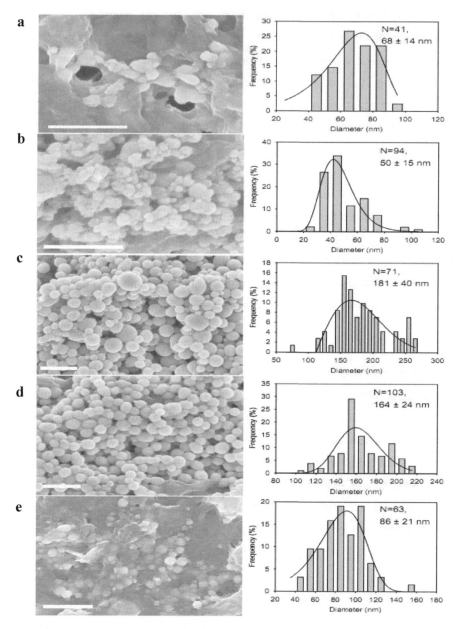

Figure 2.4. SEM images and size distribution of selenium nanoparticles produced by *Shewanella* sp HN-41 under different conditions. - N_2-purged incubations at a) 4 $^{\circ}$C , b) 15 $^{\circ}$C and c) 30 $^{\circ}$C ; d) N_2-O_2 purged incubations and e) O_2 purged incubations. N stands for the number of particles counted and the average size and standard deviations are also given. Log normal distribution is shown by solid lines (Reproduced with permission from Lee et al., 2007).

2.3.3. Structure of selenium nanoparticles

Biogenic selenium nanoparticles produced by *S.barnesii, S. shriftii* and *B. selenitireducens* have displayed features in its UV-visible spectra as compared to featureless spectra of chemically formed black selenium particles (Oremland et al., 2004). Selenium nanospheres produced from *S. shriftii* exhibited broader absorption spectra at wavelengths greater than 600 nm, indicating bimodal distribution consisting of single Se chains and polymer Se (formed after the van der Waals interaction between two or more octahedral Se rings).

Selenium nanoparticles produced by *S.barnesii, S. shriftii* and *B. selenitireducens* also exhibit Raman spectra with different features (Oremland et al., 2004). Selenium nanospheres produced by *S. barnesii* and *B. selenitireducens* formed Se_6 conformation (*i.e.* chains of 6 Se atoms), while *S. shriftii* nanoparticles had a Se_8 (*i.e.* chain of 8 Se atoms). The Raman spectra of selenium nanoparticles produced by *S. shriftii* displayed a feature at 260 cm^{-1} that indicates a single chain of Se while a feature at 234 cm^{-1} indicates Se polymer formation, thus further confirming the bimodal distribution. Selenium nanospheres formed by *S. barnesii* and *B. selenitireducens* had a Se_6 structure, but their vibrational spectra differ from each others. This is indicative that they differed in the configuration of the Se_6 chains. For selenium nanospheres produced by *B. selenitireducens*, Se_6 vibrational modes A1g and Eg were dominated in the table D_{3d} (chair) structure as compared to those in the unstable C_{2v} (boat) structure of selenium nanospheres formed by *S.barnesii*.

The spherical (50 - 400 nm) monoclinic selenium nanoparticles produced by *Bacillus subtilis* changed into an anisotropic, one-dimensional (1D) trigonal structure in 24 h when kept at ambient temperature in aqueous solution (Figure 2.5; Wang et al., 2010). The color of the solution changed from red to black that can be attributed to the formation of trigonal selenium nanowires. X-ray diffraction (XRD) analysis of these selenium nanoparticles also confirmed the transformation of monoclinic selenium nanoparticles to trigonal selenium nanowires. All peaks were in accordance to characteristic peaks of trigonal selenium. A packing of long helical chains of selenium atoms in a space group gave the cell parameters that corresponds to single phase trigonal structured selenium. Based on Raman spectra analysis and XRD, this study proposed a model for the transformation of monoclinic selenium nanoparticles to trigonal selenium nanowires (Figure 2.6). The first step of the proposed model is the reduction of selenite on the surface of proteins as the negative charge of selenite interacts with the positively charged groups of proteins. The formed selenium atoms would then act as nuclei and would grow in size following the Ostwald ripening process (Gates et al., 2002). Smaller particles dissolve or merge into larger particles which grow in size following the Gibbs - Thomson law (Elhadj et al., 2008). This study highlights the role of proteins in determining the shape of selenium nanoparticles.

A similar phenomenon, the transformation of monoclinic Se to trigonal Se during the course of incubation, was observed in selenium nanoparticles produced by *Pseudomonas alcaliphila* (Figure 2.7; Zhang et al., 2011). A peak was observed at 254 cm^{-1} in the Raman spectra after 24 h of incubation indicating the presence of monoclinic Se. However, when Raman spectra were taken after 48 h of incubation, a peak was observed at 234 cm^{-1}, which can be assigned to vibration of trigonal Se helical chains, indicating transformation of monoclinic Se to trigonal Se.

Figure 2.5. Spherical selenium nanoparticles change their crystal structure from monoclinic to trigonal selenium over time. This transformation was observed on selenium nanoparticles produced by Bacillus *subtilis*. Field emission scanning electron microscope (FESEM) and TEM images of selenium nanoparticles a) 0 h, b) 12 h, c) 24 h and d) the high magnification of (c). TEM image (e) and electron diffraction pattern (f) of an individual Se nanowire (Reproduced with permission from Wang et al., 2010).

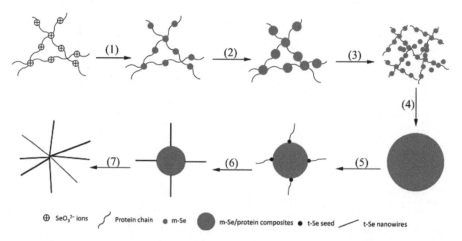

Figure 2.6. Illustration of the transformation of selenious acid to selenium nanowire. The process involved biological reduction of selenite ion by proteins produced by *B. subtilis* (1). The end product of the step 1 are selenium particles linked with proteins. More elemental selenium particles join and this complex grows in size (2). Numerous complexes of protein and elemental selenium join together to form a mesh like structure (3) which grow to become larger spherical particles (4). These spherical particles then transform to produce smaller spheres and trigonal selenium nanowires seeds (5) which eventually grow to become long trigonal selenium nanowires at the cost of spherical selenium nanoparticles (6, 7) (Reproduced with permission from Wang et al., 2010).

X-ray Absorption Fine Structure (XAFS) spectroscopy is a powerful technique that can be applied for determining the solid phase speciation of selenium in a direct manner. This technique can assess the speciation of amorphous as well as crystalline samples. X-ray Absorption Near Edge Surface (XANES) spectra of large sets of model selenium compounds were recorded in order to find out the speciation of solid phase selenium precipitated in anaerobic bioreactors (Table 2.2; van Hullenbusch et al., 2007; Lenz, 2008; Lenz, 2008c).

Figure 2.7. Transformation of red spherical elemental selenium to trigonal elemental selenium nanowires, (a) Spherical selenium particle; (b) Start of formation of selenium nanowires; (c) Formation of more nanowires and their aggregation; (d) Single nanowire (Reproduced with permission from Zhang et al., 2011).

Samples recovered from UASB reactors operated under sulfate reducing and methanogenic conditions were analyzed using XANES (Figure 2.8; van Hullenbusch et al., 2007; Lenz, 2008). On the basis of the main edge crest, spectra of both samples can be assigned to trigonal grey selenium and vitreous, black selenium. However, the first point of inflection for both samples does not match to any compound and is varied from trigonal selenium and vitreous black selenium by 0.4 and 0.5 eV. Also, a variation of only 0.3 eV was observed in both the main edge crest and the first inflection point between samples and model compounds such as ferroselite. Thus, the contribution of ferroselite in XANES spectra cannot be excluded. For the samples of a sulfate reducing bioreactor, linearly combined modeled speciation showed exclusively the contribution of trigonal grey selenium. On the other hand, samples from the methanogenic reactor consist of trigonal grey selenium and selenide. Although the main crest edge, inflection point and linear combination indicated trigonal elemental selenium, the Extended X-ray Absorption of Fine Structure Fourier Transform (EXAFS FT) did not show the second selenium neighbors in the bioreactor samples observed in the model compound (Figure 2.8). It is suggested that the selenium present in the bioreactor samples is dominantly in an aperiodic form of elemental selenium, most likely red amorphous selenium due to the visual red color and absence of XRD signal.

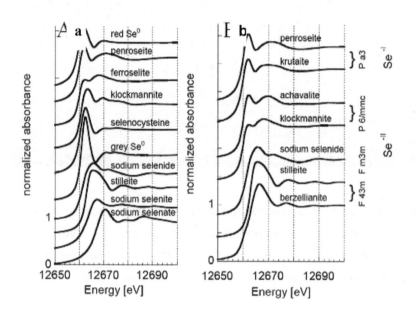

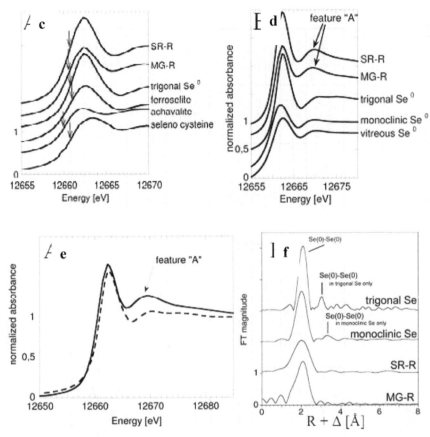

Figure 2.8. Normalized Se K-edge XANES of (a) model compounds; (b) model selenium compounds with different oxidation states and space groups; samples of UASB reactor operated under sulfate reducing conditions (SR-R) and methanogenic conditions (MG-R) with arrows point to the first inflection point (c) and feature "A" (d); (e) linear combination of model compounds (dashed line); (f) Fourier Transforms of EXAFS spectra of model compounds and samples obtained from UASB reactor operated under sulfate reducing (SR-R) and methanogenic (MG-R) conditions (Reproduced with permission from van Hullenbusch et al., 2007; Lenz, 2008).

Table 2.2. Selenium model compounds studied with main edge crest and first inflection points of the XANES spectra (Reproduced with permission from van Hullenbusch et al., 2007, Lenz, 2008).

Specimen	Chemical formula	Origin	Formal oxidation state	Crystal system	Space group*	Main edge crest (eV)*	First inflection point (eV)*
Achavalite	FeSe	Synthetic	-II	Dihexagonal Dipyramidal	P 6/mmc	12662.2	12659.9
Klockmannite	CuSe	Synthetic	-II	Dihexagonal Dipyramidal	P 6/mmc	12662.3	12660.0
Selenocysteine	$C_3H_7NO_2Se$	Synthetic	-II	NA	NA	12663.4	12660.7
Sodium selenide	Na_2Se	Synthetic	-II	Cubic	F m3m	12665.8	12661.1
Stilleite	ZnSe	Synthetic	-II	Cubic	F 43m	12665.7	12662.9
Berzelianite	Cu_2Se	Czech Republic	-II	Cubic	F 43m	12666.6	12663.0
Berzelianite	Cu_2Se	Sweden	-II	Cubic	F 43m	12666.9	12662.9
Penroseite	$(Ni, Co, Cu)Se_2$	Bolivia	-I	Isometric - Diploidal	P a3	12662.1	12659.9
Krutaite	$CuSe_2$	Bolivia	-I	Isometric - Diploidal	P a3	12662.1	12659.8
Ferroselite	$FeSe_2$	Utah (USA)	-I	Isometric - Dipyramidal	P nnm	12662.8	12660.9
Red α-monoclinic Se(0)	Se	Synthetic	0	Monoclinic	P 2/n	12662.1	12659.5
Black, vitreous Se(0)	Se	Synthetic	0	Amorphous	NA	12662.5	12660.0
Grey Se(0)	Se	New Mexico (USA)	0	Trigonal	P 321	12662.5	12661.1
Sodium selenite	Na_2SeO_3	Synthetic	+IV	Monoclinic	P 2/c	12667.3	12664.1
Sodium selenate	Na_2SeO_4	Synthetic	+VI	Orthorhombic	P ddd	12670.8	12667.9
Sample from sulfate reducing reactor	TBD	Sulphate reducing reactor	**TBD**	**TBD**	**TBD**	12662.5	12660.6
Sample from methanogenic reactor	TBD	Methanogenic reactor	**TBD**	**TBD**	**TBD**	12662.5	12660.5

TBD - To be determined; NA - Not applicable

* For more detailed explanation of space groups, please refer to Ladd, 2003.

XANES has also been used for scrutinizing the selectivity of the chemical extraction methods towards the speciation of selenium (Lenz et al., 2008c). It was observed that during the chemical extraction method, 58% selenium that is present as metal selenides and organic selenium compounds is estimated as the elemental selenium fraction. In this study, the best fit for the selenium precipitation in selenate treating UASB anaerobic granules was obtained using four model compounds (Figure 2.9). Out of these four model compounds, two model compounds were dominated in both anaerobic and aerobic (10 minutes exposure to air) extraction. One of the dominant compounds in both the aerobic and anaerobic extraction was trigonal elemental selenium. In the case of anaerobic extraction, the other dominant compound was stilleite or sodium selenide. In the case of aerobic extraction, stilleite or achavalite was the other dominant form. The effect of short exposure of air during the sequential extraction procedure was also investigated using XANES. The presence of highly oxidized species in the first (extraction with 0.25 M KCl) and second step (extraction with 0.1 M K_2HPO_4) of the aerobic sequential extraction procedure after 10 minutes exposure to air can be attributed to oxidation of organic selenocysteine like species. Change in speciation after the third step of the aerobic as compared to the anaerobic sequential extraction procedure (extraction with 0.25 M Na_2SO_3, sonication at 20 kHz for 2 min, then ultrasonic bath for 4 h) was attributed to the oxidation of cubic (sodium selenide, -II) type to elemental selenium. However, the cubic compound is more likely to be an insoluble PbSe type compound because sodium selenide is soluble and highly labile.

The selenium oxidation state in the selenium nanoparticles was examined using XANES (Lee et al., 2007). Selenium K-edges XANES spectra for biogenic produced selenium by *Shewanella* sp. HN-41 and four other model compounds depicting -2 (FeSe as selenide), 0 (elemental selenium powder), +4 (sodium selenite) and +6 (sodium selenate) was obtained (Figure 2.10a). The first feature in the normalized absorbance of XANES spectra of biogenic selenium nanoparticles cannot be used to distinguish between the elemental selenium and selenide. To get more insight, first inflection point energies of absorption edges were examined (2.10b). When comparing the inflection point energies of biogenic selenium nanoparticles with those of elemental selenium powder and selenide (FeSe), the inflection point energies of biogenic selenium nanoparticles was closer to elemental selenium than that of selenide.

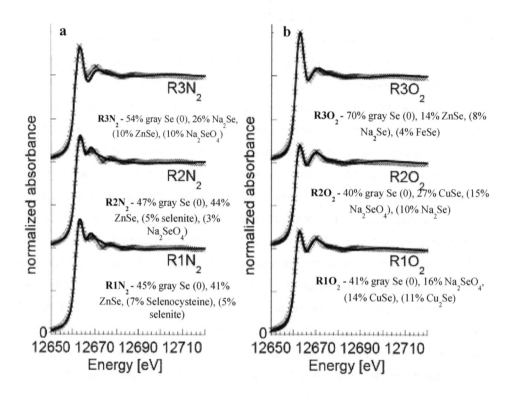

Figure 2.9. Normalized Se K-edge XANES spectra for sequential extraction residues R1, R2 and R3 to R3 (solid lines) and best fit by linear combination of model compounds (×) after extraction performed anoxically (a) and under ambient air (b). Contributions by model compounds (text box in the chart) to the best fit results are given in % relations. Misfits are related to unidentified selenium species (Lenz et al., 2008c). R1, R2 and R3 are defined as the residual obtained after the first, second and third of sequential extraction procedure, respectively (Reproduced with permission from Lenz et al., 2008c).

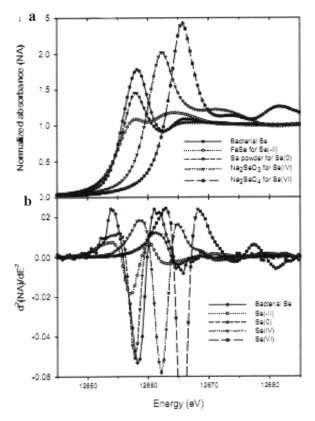

Figure 2.10. Normalized Se K-edge (a) XANES spectra and (b) the second derivatives for elemental selenium nanoparticles produced by *Shewanella* sp. HN-41 and model compounds (FeSe, Se powder, Na_2SeO_4 and Na_2SeO_3) (Reproduced with permission from Lee et al., 2007).

2.4. Challenges in biogenic selenium nanoparticles production

The biogenic production of selenium nanoparticles by reduction of selenium oxyanions is a bottom up process that follows the Ostwald ripening principle and thus the size of the formed selenium nanoparticles increases with time (Zhang et al., 2011). One of the most important challenges in the biological production of selenium nanoparticles is being able to control the size and polydispersity index of the particles. The importance of controlling the size lies in the fact that many properties of nanoparticles such as optoelectronic, material and catalytic properties are affected by the size of the nanoparticles. The biologically produced selenium nanoparticles

33

are polydisperse with an average diameter greater than 100 nm (Oremland et al., 2004; Tejo Prakash et al., 2009; Dhanjal and Cameotra, 2010; Kuroda et al., 2011; Zhang et al., 2011; Bajaj et al., 2012). Since current nanotechnology applications use particles much smaller than 100 nm, there is a need to understand the mechanisms of biogenic formation of selenium nanoparticles so that effective control of their size can be achieved.

Another challenge in the biogenic production of selenium nanoparticles is their purification. As selenium nanoparticles can also be formed intracellularly, the separation of these particles from the biomass without altering their properties is extremely challenging. The ideal situation would be that selenium nanoparticles are produced extracellularly.

2.4.1. Controlling the size of selenium nanoparticles

Selection of an appropriate capping agent can control the size and shape of the nanoparticles (Pramanik et al., 2007; Saraswathi et al., 2007; Lu et al., 2008; Li et al., 2013). Biomacro-molecules such as proteins and DNA act as capping agent by attaching to nanoparticles, thus preventing uncontrolled growth and limiting the size of nanoparticles (Niemeyer, 2001). A study by Dobias et al. (2011) showed the role of proteins in controlling the shape and size distribution of biogenic selenium nanoparticles. Cell free extract of *E. coli* grown in the presence of selenite was used to expose biologically produced selenium nanoparticles and iron nanoparticles. Chemically produced selenium nanoparticles were also produced in the presence of cell free extracts. It was found that chemically produced selenium nanoparticles in presence of cell free extract showed a more narrow size distribution (106.7 ± 8.7 nm) in comparison to chemically produced selenium nanoparticles in the absence of cell free extracts (10 - 90 nm). This suggests that the cell free extract stabilizes the nanoparticles or provides a template for the controlled growth of crystals. Six different proteins were found to bind to selenium nanoparticles. Two of these proteins (EF-Tu and 3-oxoacyl synthase) were found to bind non-specifically as these two proteins could bind to iron nanoparticles as well. The other four proteins were found to be specifically and strongly attached to selenium nanoparticles. These four proteins were isocitrate lyase, isocitrate dehydrogenase, outer membrane

34

protein C precursor and alcohol dehydrogenase. All these four proteins have a size in the range of 36 to 48 KDa, exhibit enzymatic to structural functions and have an isoelectric point between pH 4.58 to 5.94. The authors also demonstrated that when the chemical synthesis of selenium nanoparticles occurs in the presence of alcohol dehydrogenase, the size of the produced nanoparticles was three fold smaller (122 ± 24 nm) than the size of chemically synthesized selenium nanoparticles in absence of alcohol dehydrogenase (319 ± 57 nm).

2.4.2. Controlling the location of selenium nanoparticle production - intra or extracellular

Biogenic selenium nanoparticles can be produced either extracellularly or intracellularly. One of the proposed mechanisms of extracellular production is reduction of the selenium oxyanion via outer membrane cytochromes. For intracellular production, it has been proposed that selenium oxyanion reacts with thiols inside the cell to produce selenium nanoparticles which may be expelled outside the cell (Figure 2.11; Zannoni et al., 2008; Pierce et al., 2009). The extracellular production would give a higher yield while the intracellular production provides monodisperse nanoparticles with a better control of size.

In the study by Oremland et al. (2004), when *B. selenitireducens* was grown with nitrate, followed by a washing step and then fed with selenite, the external selenium nanospheres were predominantly produced. In another study to understand the excretion of selenium nanoparticles from inside the cell, a new protein of approximately 95 kDa was discovered. This protein was associated with selenium nanoparticles and was produced during selenate respiration by *Thauera selenatis* (Debieux et al., 2011). The protein was named Se factor A. Subsequent experiments revealed that the protein is secreted in response to increasing selenite concentration and hence is up-regulated. The genome analysis of *T. selenatis* disclosed an open reading frame that leads to a protein with an estimated mass of 94.5 kDa. Due to the absence of a cleavable signal peptide, it was suggested that the protein is exported directly from the cytoplasm. It has been demonstrated that *in vitro* production of selenium nanoparticles by reduction of selenite by glutathione (GSH, glutamylcysteinylglycine) are stabilized by the presence of Se factor A. This study also proposes a selenate reduction pathway in *T. selenatis*. In this proposed pathway, selenium nanoparticles are produced and stabilized in cytoplasm before being expelled outside the cell.

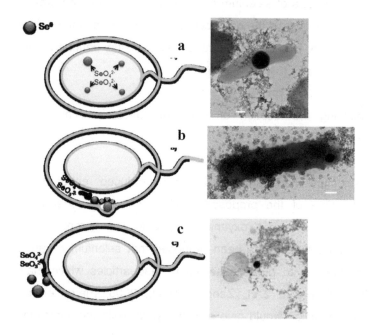

Figure 2.11. Proposed mechanisms for the biogenesis of Se^0 nanoparticles. (a) Se oxyanions react intracellularly with thiols and nanoparticles are produced (Reproduced with permission from Zannoni et al., 2008. (b) Se oxyanions are reduced in the periplasmic space of gram-negative bacteria; Se^0 may be excreted via vesiculation. (c) Reduction via outer membrane cytochromes (Reproduced with permission from Pearce et al., 2009. Transmission electron microscopic (TEM) images are from Gonzalez-Gil et al., in preparation) showing anaerobic granular sludge contains various microorganisms that can synthesize Se^0 nanoparticles. Scale bars, 0.2 μm.

2.4.3. Role of synthetic biology in achieving biogenic production of selenium nanoparticles

The biogenic production of selenium nanoparticles involves "a bottom-up approach" meaning that a single atom joins together with other atoms or molecules to form nanoparticles. This growth process is affected by the presence of organic molecules such as proteins, DNA and sugars. These molecules act as "templates" for nucleation and control the shape and size of the resulting crystals (Niemeyer, 2001). The growth process is also affected by the concentration of the solute and the temperature of the system. Thus, it is important to not only understand the complete mechanism and factors affecting the formation of selenium nanoparticles, but also to have the ability to regulate the concentration of proteins and reducing agents involved in their production. For example, a mutant of *Shewanella oneidensis* lacking the outer membrane C type cytochromes produced smaller size Ag and AgS nanoparticles as compared to *Shewanella oneidensis* with outer membrane C type cytochromes (Ng et al., 2013). Moreover, the nanoparticles produced by the mutant showed higher antibacterial and catalytic activities. Thus, by playing with the expression profile of important proteins, the size and activity of nanoparticles could be better controlled.

The chemical reduction of sodium selenite by glutathione (GSH, glutamylcysteinylglycine) in the presence of bovine serum albumin at room temperature resulted in the production of particles smaller than 100 nm (Johnson et al., 2008). This process closely resembles the dissimilatory reduction of sodium selenite in *R. rubrum* and *E. coli* (Kessi and Hanselmann, 2004). In *R. rubrum*, the selenite reduction rate decreased with decreasing glutathione (GSH, glutamylcysteinylglycine) concentration, whereas in *E. coli*, the synthesis of glutathione (GSH, glutamylcysteinylglycine) was induced when selenite was present. This suggests that by appropriately controlling the expression of glutathione (GSH, glutamylcysteinylglycine) and by eliminating other factors such as expression of any other reductase or excess production of any other protein that may impact the crystal growth, it may be possible to produce selenium nanoparticles extracellularly with *E. coli* produced glutathione (GSH, glutamylcysteinylglycine). To achieve this objective, a detailed understanding of cell survivability, cell growth, reduction pathways and mechanisms is required. Once we have this understanding, a synthetic cell with minimal functions can be designed where expression of every protein is tightly regulated to produce selenium nanoparticles with desired characteristics.

The Se factor A protein found in *T. selenatis* stabilizes the selenium nanoparticles inside the cell prior to be expelled outside the cell (Debieux et al., 2011). However, the transport mechanism including trigger factors for transporting selenium nanoparticles from inside the cell to outside is not understood. A detailed understanding of this transport mechanism can help to trigger expulsion of selenium nanoparticles when they have reached a desired size.

The role of enzymes and proteins in the production of selenium nanoparticles and their stability is well known (Kessi and Hanselmann, 2004; Yee et al., 2007; Ma et al., 2009; Choudhury et al., 2011; Lenz et al., 2011). However, till to date only the structure of selenate reductase in *Thauera selenatis* has been studied in detail (Maher and Joan, 2006; Dridge et al., 2007). Our understanding of the regulation, structure and active sites of proteins involved in the selenate or selenite reduction is severely lacking. The expression levels of these proteins determine the rate of reaction and hence affect the growth of selenium nanoparticles. The active sites in the 3-dimensional (3D) structure of the protein can act as a template for controlling the growth of selenium nanoparticles. A better understanding of the expression levels and the 3D structure of these proteins would lead to better control of size and polydispersity of biologically produced selenium nanoparticles.

The understanding of complete mechanism and pathways of biological reduction of selenium oxyanions as well as the need of designing new synthetic micro-organism can be achieved by synthetic biology approach. The power of synthetic biology lies in combining the knowledge of metagenomics, proteomics, structural biology, molecular biology and bioinformatics (Marner, 2009; Ellis and Goodacre, 2012; Lam et al., 2012; Velenich and Gore, 2012; Lim et al., 2013). After the demonstration of the first devices in 2000 – the genetic toggle switch and the genetic oscillator – synthetic biology has grown rapidly from single gene/protein devices to more complex transcription and signaling networks (Elowitz and Leibler, 2000; Gardner et al., 2000). However, the beauty of synthetic biology lies in its engineering-driven approaches of modularization, rationalization and modeling. It can play a role in analyzing and synthesizing the signaling pathways and cellular control (Marner, 2009. Synthetic biology can help us in reconstruction of the natural pathways in evolutionary distant hosts. This will help in subtracting the interference from the original host which in turn will help us in developing strategies in designing more complex signaling networks at DNA, RNA and protein level to reprogram cellular functions for selenium nanoparticles production (Figure 2.12; Deplazes, 2009).

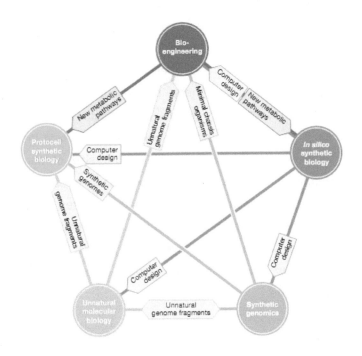

Figure 2.12. Illustration of different categories of synthetic biology. In silico design would help in minimizing the experiments and is thus useful for all parts of synthetic biology. Synthetic genomes can be used in designing model organisms with completely synthetic genome. For example, a synthetic genome encoding new metabolic pathways can be integrated in protocells for the production of selenium nanoparticles with optimized desired characteristics. The same approach can be used for understanding the mechanism of selenium oxyanions reduction by transferring the natural pathway in evolutionary distant host (Reproduced with permission from Deplazes, 2009).

2.5. Conclusions

The biogenic production of selenium nanoparticles is a promising eco-friendly option to produce selenium nanoparticles at ambient temperature and pressure. Biological production methods also give cost advantage vis a vis chemical methods as they do not require the use of specialized equipment and costly chemicals. The major challenges in the biological production method are poor product quality (higher polydispersity and larger size) and need for exhaustive post production treatment. However, there are enough evidences present in the literature to suggest that the above challenges can be addressed provided that we understand the detailed mechanisms involved in the biogenic formation of selenium nanoparticles. This understanding would allow researchers to optimize the presently known microorganisms and or to completely design a new synthetic microorganism with desired properties. These desired properties will include inducing the extracellular production of selenium nanoparticles, controlling the size of nanoparticles by controlling the expression of desired reducing agents and expression of appropriate capping agents in the exact amount that would lead to desired surface properties, size and monodispersity. The expression of an appropriate template can also lead to the formation of smart self assembled systems. Synthetic biology has a lot of potential to completely revolutionize the biogenic production of selenium nanoparticles in near future.

Acknowledgement

The authors thank the EU for providing financial support through the Erasmus Mundus Joint Doctorate Programme ETeCoS[3] (Environmental Technologies for Contaminated Solids, Soils and Sediments, grant agreement FPA n°2010-0009.

2.6 References

Abdelouas, A., Gong, W.L., Lutze, W., Shelnutt, J.A., Franco, R. and Moura, I., 2000. Using Cytochrome C$_3$ to make selenium nanowires. Chem. Matter. 12, 1510–1512.

Astratinei, V., van Hullebusch, E. and Lens, P. 2006. Bioconversion of selenate in methanogenic anaerobic granular sludge. J Env. Qual. 35, 1873–83.

Bajaj, M., Schmidt, S. and Winter, J., 2012. Formation of Se (0) nanoparticles by *Duganella* sp. and *Agrobacterium* sp. isolated from Se-laden soil of North-East Punjab, India. Microb. Cell Fact. 11, 64.

Belzile, N., Wu, G.J., Chen, Y.W. and Appanna, V.D., 2006. Detoxification of selenite and mercury by reduction and mutual protection in the assimilation of both elements by *Pseudomonas fluorescens*. Sci. Total Environ. 367, 704-714.

Chen, H., Shin, D., Nam, J., Kwon, K., Yoo, J., 2010. Selenium nanowires and nanotubes synthesized via a facile template-free solution method. Mater. Res. Bull. 45, 699–704.

Choudhury, H.G., Cameron, A.D., Iwata, S., Beis, K., 2011. Structure and mechanism of the chalcogen-detoxifying protein TehB from *Escherichia coli*. Biochem. J. 435, 85–91.

Deplazes, A., 2009. Piecing together a puzzle. EMBO Reports. 10, 428–432.

Debieux, C.M., Dridge, E.J., Mueller, C.M., Splatt, P., Paszkiewicz, K. and Knight, I., 2011. A bacterial process for selenium nanosphere assembly. PNAS. 108, 13480–13485.

Dhanjal, S., Cameotra, S.S., 2010. Aerobic biogenesis of selenium nanospheres by *Bacillus cereus* isolated from coalmine soil. Microb. Cell Fact. 9, 52.

Dobias, J; Suvorova, E.I, Bernier-latmani, R., 2011. Role of proteins in controlling selenium nanoparticle size. Nanotechnology 22, 195605. Dridge.

Dridge, E.J., Watts, C.A., Jepson, B.J.N., Line, K., Santini, J.M. and Richardson, D.J., 2007. Investigation of the redox centres of periplasmic selenate reductase from *Thauera selenatis* by EPR spectroscopy. Biochem. J. 408(1), 19–28

Dwivedi, C., Shah, C.P., Singh, K., Kumar, M., Bajaj, P.N., 2011. An Organic Acid-induced Synthesis and Characterization of Selenium Nanoparticles. J. Nanotechnol. 2011, 1–6.

Elhadj, S., Chernov, A.A. and Yoreo, J.J.D., 2008. Solvent-mediated repair and patterning of surfaces by AFM. Nanotechnology. 19, 105304–105312.

Ellis, D.I., Goodacre, R., 2012. Metabolomics-assisted synthetic biology. Curr. Opin. Biotechnol. 23, 22–28.

Elowitz, M.B., Leibler, S., 2000. A synthetic oscillatory network of transcriptional regulators. Nature. 403, 335–338.

European Commission Recommendation 2011/696/EU, OJ L 275, (October 20, 2011).

Fellowes, J.W., Pattrick, R.A.D., Green, D.I., Dent, A., Lloyd, J.R., Pearce, C.I., 2011. Use of biogenic and abiotic elemental selenium nanospheres to sequester elemental mercury released from mercury contaminated museum specimens. J. Hazard. Mater. 189, 660–669.

Fesharaki, P.J., Nazari, P., Shakibaie, M., Rezaie, S., Banoee, M., Abdollahi, M., Shahverdi, A.R., 2010. Biosynthesis of selenium nanoparticles using Klebsiella pneumoniae and their recovery by a simple sterilization process. Brazilian J. Microbiol. 41, 461–466.

Gardner, T.S., Cantor, C.R., Collins, J.J., 2000. Construction of a genetic toggle switch in Escherichia coli. Nature. 403, 339–342

Gates, B., Mayers, B., Cattle, B., Xia, Y., 2002. Synthesis and characterization of uniform nanowires of trigonal Selenium. Adv. Funct. Mater. 12, 219–227.

Gharieb, M.M., 1995. Reduction of selenium oxyanions by unicellular, polymorphic and filamentous fungi: cellular location of reduced selenium and implications for tolerance. J. Ind. Microbiol. 14, 300–311.

Hunter, W.J., 2007. An Azospira oryzae (syn Dechlorosoma suillum) strain that reduces selenate and selenite to elemental red selenium. Curr. Microbiol. 54, 376–381.

Johnson, N.C., Manchester, S., Sarin, L., Gao, Y., Kulaots, I., Hurt, R.H., 2008. Mercury vapor release from broken compact fluorescent lamps and in situ capture by new nanomaterial sorbents. Environ. Sci. Technol. 42, 5772–5778.

Kessi, J., Hanselmann, K.W., 2004. Similarities between the abiotic reduction of selenite with glutathione and the dissimilatory reaction mediated by Rhodospirillum rubrum and Escherichia coli. J. Biol. Chem. 279, 50662–50669.

Kessi, J., Ramuz, M., Wehrli, E., Spycher, M. and Bachofen, R., 1999. Reduction of Selenite and Detoxification of Elemental Selenium by the Phototrophic Bacterium Rhodospirillum rubrum. Appl. Environ. Microbiol. 65, 4734-40

Klonowska, A., Heulin, T., 2005. Selenite and Tellurite Reduction by Shewanella oneidensis. Appl. Environ. Microbiol.71, 5607–5609.

Kuroda, M., Notaguchi, E., Sato, A., Yoshioka, M., Hasegawa, A., Kagami, T., Narita, T., Yamashita, M., Sei, K., Soda, S., Ike, M., 2011. Characterization of Pseudomonas stutzeri NT-I capable of removing soluble selenium from the aqueous phase under aerobic conditions. J. Biosci. Bioeng. 112, 259–64.

Ladd, M.F.C. and Palmer, L.A., 2003. Structure determination by X-ray crystallography, fourth ed. Kluwer Academic/Plenum Publishers, New York

Lam, C.M.C, Diez, M.S., Godinho, M. and Martins dos Santos, V.A.P., 2012. Programmable bacterial catalysis - designing cells for biosynthesis of value-added compounds. FEBS Lett. 586, 2184–2190.

Langi, B., Shah, C., Singh, K., Chaskar, A., Kumar, M., 2010. Ionic liquid-induced synthesis of selenium nanoparticles. Mater. Res. Bull. 45, 668–671.

Lee, J.-H., Han, J., Choi, H., Hur, H.-G., 2007. Effects of temperature and dissolved oxygen on Se(IV) removal and Se(0) precipitation by *Shewanella* sp. HN-41. Chemosphere 68, 1898–1905.

Lenz, M., 2008. Biological selenium removal from wastewater. Ph.D. Thesis. Wageningen University, The Netherlands, 3, 37–46.

Lenz, M., Hullebusch, E.D. Van, Hommes, G., Corvini, P.F.X., Lens, P.N.L., 2008a. Selenate removal in methanogenic and sulfate-reducing upflow anaerobic sludge bed reactors. Water Res. 42, 2184–2194.

Lenz, M., Kolvenbach, B., Gygax, B., Moes, S., Corvini, P.F.X., 2011. Shedding light on selenium biomineralization: proteins associated with bionanominerals. Appl. Environ. Microbiol. 77, 4676–4680.

Lenz, M., Lens, P.N.L., 2009. The essential toxin: the changing perception of selenium in environmental sciences. Sci. Total Environ. 407, 3620–33.

Lenz, M., Smit, M., Binder, P., van Aelst, A.C., Lens, P.N.L., 2008b. Biological alkylation and colloid formation of selenium in methanogenic UASB reactors. J. Environ. Qual. 37, 1691–700.

Lenz, M., van Hullebusch, E.D., Farges, F., Nikitenko, S., Borca, C.N., Grolimund, D., Lens, P.N.L., 2008c. Selenium speciation assessed by X-ray absorption spectroscopy of sequentially extracted anaerobic biofilms. Environ. Sci. Technol. 42, 7587–7593.

Lenz, M., van Hullebusch, E.D., Farges, F., Nikitenko, S., Corvini, P.F.X., Lens, P.N.L., 2011a. Combined speciation analysis by X-ray absorption near-edge structure spectroscopy, ion chromatography, and solid-phase microextraction gas chromatography-mass spectrometry to evaluate biotreatment of concentrated selenium wastewaters. Environ. Sci. Technol. 45, 1067–73.

Li, C.-C., Chang, S.-J., Su, F.-J., Lin, S.-W., Chou, Y.-C., 2013. Effects of capping agents on the dispersion of silver nanoparticles. Colloids Surfaces A Physicochem. Eng. Asp. 419, 209–215.

Lim, W.A, Lee, C.M. and Tang, C., 2013. Design principles of regulatory networks: searching for the molecular algorithms of the cell. Mol. Cell 49, 202–212.

Lortie, L., Gould, W.D., Rajan, S., McCready, R.G., Cheng, K.J., 1992. Reduction of Selenate and Selenite to Elemental Selenium by a Pseudomonas stutzeri Isolate. Appl. Environ. Microbiol. 58, 4042–4044.

Losi, M.E., Frankenberger, W.T., 1997. Reduction of Selenium Oxyanions by Enterobacter cloacae SLD1a-1: Isolation and Growth of the Bacterium and Its Expulsion of Selenium Particles. Appl. Environ. Microbiol. 63, 3079–3084.

Lu, Y., Lu, X., Mayers, B.T., Herricks, T., Xia, Y., 2008. Synthesis and characterization of magnetic Co nanoparticles: A comparison study of three different capping surfactants. J. Solid State Chem. 181, 1530–1538.

Ma, J., Kobayashi, D.Y., Yee, N., 2009. Role of menaquinone biosynthesis genes in selenate reduction by *Enterobacter cloacae* SLD1a-1 and *Escherichia coli* K12. Environ. Microbiol. 11, 149–158.

Ma, J., Liu, X., Wu, Y., Peng, P., Zheng, W., 2008. Controlled synthesis of selenium of different morphologies at room temperature. Cryst. Res. Technol. 43, 1052–1056.

Maher, M.J., Joan, M., 2006. crystallization papers Crystallization and preliminary X-ray analysis of the selenate reductase from *Thauera selenatis* crystallization papers 706–708.

Marner, W.D., 2009. Practical application of synthetic biology principles. Biotechnol. J. 4, 1406–1419.

Mishra, R.R., Prajapati, S., Das, J., Dangar, T.K., Das, N., Thatoi, H., 2011. Reduction of selenite to red elemental selenium by moderately halotolerant *Bacillus megaterium* strains isolated from Bhitarkanika mangrove soil and characterization of reduced product. Chemosphere 84, 1231–1237.

Ng, C.K., Sivakumar, K., Liu, X., Madhaiyan, M., Ji, L., Yang, L., Tang, C., Song, H., Kjelleberg, S., Cao, B., 2013. Influence of outer membrane c-type cytochromes on particle size and activity of extracellular nanoparticles produced by *Shewanella oneidensis*. Biotechnol. Bioeng. 110, 1831–1837.

Niemeyer, C.M., 2001. Nanoparticles, proteins, and nucleic Acids, Biotechnology meets materials Science. Angew. Chem. Int. Ed., 40, 4128–4158.

Oremland, R.S., Herbel, M.J., Blum, J.S., Langley, S., Beveridge, T.J., Ajayan, P.M., Sutto, T., Ellis, A. V, Curran, S., 2004. Structural and spectral features of selenium nanospheres produced by Se-respiring bacteria. Appl. Environ. Microbiol. 70, 52–60.

Pearce, C.I., Coker, V.S., Charnock, J.M., Pattrick, R. a D., Mosselmans, J.F.W., Law, N., Beveridge, T.J., Lloyd, J.R., 2008. Microbial manufacture of chalcogenide-based nanoparticles via the reduction of selenite using Veillonella atypica: an in situ EXAFS study. Nanotechnology 19, 155603.

Pearce, C.I., Pattrick, R. a D., Law, N., Charnock, J.M., Coker, V.S., Fellowes, J.W., Oremland, R.S., Lloyd, J.R., 2009. Investigating different mechanisms for biogenic selenite transformations: Geobacter sulfurreducens, Shewanella oneidensis and Veillonella atypica. Environ. Technol. 30, 1313–1326.

Pramanik, N., Tarafdar, A., Pramanik, P., 2007. Capping agent-assisted synthesis of nanosized hydroxyapatite: Comparative studies of their physicochemical properties. J. Mater. Process. Technol. 184, 131–138.

Qu, X., Alvarez, P.J.J., Li, Q., 2013. Applications of Nanotechnology in Water and Wastewater Treatment. Water Res. 47, 3931–3946

Saraswathi Amma, B., Ramakrishna, K., Pattabi, M., 2007. Comparison of various organic stabilizers as capping agents for CdS nanoparticles synthesis. J. Mater. Sci. Mater. Electron. 18, 1109–1113.

Shah, C.P., Dwivedi, C., Singh, K.K., Kumar, M., Bajaj, P.N., 2010. Riley oxidation: A forgotten name reaction for synthesis of selenium nanoparticles. Mater. Res. Bull. 45, 1213–1217.

Shah, C.P., Singh, K.K., Kumar, M., Bajaj, P.N., 2010. Vinyl monomers-induced synthesis of polyvinyl alcohol-stabilized selenium nanoparticles. Mater. Res. Bull. 45, 56–62.

Shahverdi, A.R., Fakhimi, A., Mosavat, G., Jafari-Fesharaki, P., Rezaie, S. and Rezayat, S.M., 2010. Antifungal activity of biogenic selenium nanoparticles. World Appl. Sci. J. 10, 918–922.

Smith, T.W. and Cheatham, R.A., 1980. Functional polymers in the generation of colloidal dispersions of amorphous selenium. Macromolecules, 13, 1203–1207.

Stolz, J.F., Basu, P., Santini, J.M., Oremland, R.S., 2006. Arsenic and selenium in microbial metabolism. Annu. Rev. Microbiol. 60, 107–130.

Switzer, J., Allana, B., Bindi, B., Buzzelli, J., Stolz, J.F., Oremland, R.S., 1998. and *Bacillus selenitireducens* , sp . nov .: two haloalkaliphiles from Mono Lake , California that respire oxyanions of selenium and arsenic. Arch. Microbiol. 171, 19–30.

Tam, K., Ho, C.T., Lee, J.-H., Lai, M., Chang, C.H., Rheem, Y., Chen, W., Hur, H.-G., Myung, N. V., 2010. Growth mechanism of amorphous selenium nanoparticles synthesized by *Shewanella* sp. HN-41. Biosci. Biotechnol. Biochem. 74, 696–700.

Tejo Prakash, N., Sharma, N., Prakash, R., Raina, K.K., Fellowes, J., Pearce, C.I., Lloyd, J.R., Pattrick, R. a D., 2009. Aerobic microbial manufacture of nanoscale selenium: exploiting nature's bio-nanomineralization potential. Biotechnol. Lett. 31, 1857–1862.

van Hullenbusch, E., Farges, F., Lenz, M., Lens, P., Brown Jr G.E., 2007. Selenium speciation in biofilms from granular sludge bed reactors used for wastewater treatment. AIP Conference Proceedings, 882, 229-231.

Velenich, A., Gore, J., 2012. Synthetic approaches to understanding biological constraints. Curr. Opin. Chem. Biol. 16, 323–328.

Walsh, S., Balbus, J.M., Denison, R., Florini, K., 2008. Nanotechnology: getting it right the first time. J. Clean. Prod. 16, 1018–1020.

Wang, T., Yang, L., Zhang, B., Liu, J., 2010. Extracellular biosynthesis and transformation of selenium nanoparticles and application in H_2O_2 biosensor. Colloids Surf. B. Biointerfaces 80, 94–102.

Webster, T.J., 2008. Enhanced osteoblast adhesion on nanostructured selenium compacts for anti-cancer orthopedic applications. IEEE 33rd Annual Northeast Bioengineering Conference, NEBC '07, 241-242

Yadav, V., Sharma, N., Prakash, R., Raina, K.K., Bharadwaj, L.M. and Prakash, T., 2008. Generation of selenium containing Nano-Structures by soil bacterium *Pseudomonas aeruginosa*. Biotechnology 7, 299–304.

Yamada, A., Miyashita, M., Inoue, K. and Matsunaga, T., 1997. Extracellular reduction of selenite by a novel marine photosynthetic bacterium. Appl. Microbiol. Biotechnol. 48, 367–372.

Yee, N., Ma, J., Dalia, A., Boonfueng, T. and Kobayashi, D.Y., 2007. Se(VI) reduction and the precipitation of Se(0) by the facultative bacterium *Enterobacter cloacae* SLD1a-1 are regulated by FNR. Appl. Environ. Microbiol. 73, 1914–20.

Zannoni, D., Borsetti, F., Harrison, J.J., Turner, R.J., 2008. The bacterial response to the chalcogen metalloids Se and Te., Adv. Microb. Physiol. 53, 1–71.

Zhang, W., Chen, Z., Liu, H., Zhang, L., Gao, P., Li, D., 2011. Biosynthesis and structural characteristics of selenium nanoparticles by *Pseudomonas alcaliphila*. Colloids Surf. B. Biointerfaces 88, 196–201.

Zhang, Y., Wang, J., Zhang, L., 2010. Creation of highly stable selenium nanoparticles capped with hyperbranched polysaccharide in water. Langmuir 26, 17617–23.

CHAPTER 3

Extracellular polymeric substances (EPS) govern surface charge of biogenic elemental selenium nanoparticles (BioSeNPs)

This chapter has been accepted as:

Jain, R., Jordan, N., Weiss, S., Foerstendorf, H., Heim, K., Kacker, R., Hübner, R., Kramer, H., Hullebusch, E.D. Van, Farges, F., Lens, P.N.L., 2015. Extracellular polymeric substances (EPS) govern the surface charge of biogenic elemental selenium nanoparticles (BioSeNPs). Environ. Sci. Tech. 49, 1713-1720

Abstract:

The origin of the organic layer covering colloidal biogenic elemental selenium nanoparticles (BioSeNPs) is not known, particularly in the case when they are synthesized by complex microbial communities. This study investigated the presence of extracellular polymeric substances (EPS) on BioSeNPs. The role of EPS in capping the extracellularly available BioSeNPs was also examined. FT-IR spectroscopy and colorimetric measurements confirmed the presence of functional groups characteristic of proteins and carbohydrates on the BioSeNPs, suggesting the presence of EPS. Chemical synthesis of elemental selenium nanoparticles in the presence of EPS, extracted from selenite fed anaerobic granular sludge, yielded stable colloidal spherical selenium nanoparticles. Furthermore, extracted EPS, BioSeNPs and chemically synthesized EPS capped selenium nanoparticles had similar surface properties, as shown by ζ-potential versus pH profiles and iso-electric point measurements. This study shows that the EPS of anaerobic granular sludge forms the organic layer present on the BioSeNPs synthesized by these granules. The EPS also govern the surface charge of these BioSeNPs, thereby contributing to their colloidal properties, hence affecting their fate in the environment and the efficiency of bioremediation technologies.

Keywords: EPS, BioSeNPs, surface charge, capping, FT-IR

Graphical abstract:

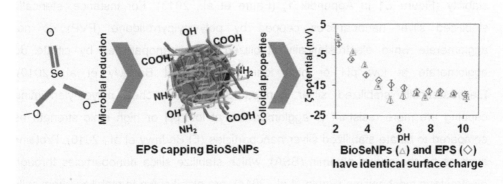

EPS capping BioSeNPs   BioSeNPs ($\triangle$) and EPS ($\lozenge$) have identical surface charge

3.1. Introduction

Selenium is an essential nutrient in the human diet (Rayman, 2000). However, the higher concentrations of selenium, especially those of the selenium oxyanions selenate and selenite, are toxic to humans, animals and aquatic life (Hamilton, 2004; Lenz and Lens, 2009; Qin et al., 2013). Therefore, regulatory agencies have set limits on total selenium discharges, e.g. the Environmental Protection Agency of the United States has recommended a discharge limit of 5 µg L^{-1} total selenium in freshwater (USEPA, 2013). Anaerobic bioreduction of dissolved selenium oxyanions to elemental selenium is considered a promising technology for the remediation of selenium oxyanions containing wastewaters (Cantafio et al., 1996; Lenz et al., 2008). However, the produced biogenic elemental selenium is in the form of colloidal spherical nanoparticles with a diameter of 50 - 500 nm (Oremland et al., 2004; Jain et al., 2014). Such colloidal biogenic elemental selenium nanoparticles (BioSeNPs) are present in high concentrations in the effluent of upflow anaerobic sludge blanket reactors (UASB), in which anaerobic granules treat selenium rich wastewaters (Lenz et al., 2008). Buchs et al. (2013) showed that the colloidal properties of these BioSeNPs determine their transport and fate in the environment as well as the bioremediation efficiency. Thus, it is important to understand the factors governing the colloidal properties of BioSeNPs.

Capping agents are known to affect the surface properties of chemically produced metal(loid) nanoparticles, including surface charge and consequently colloidal stability (Figure S1 in Appendix 1) (Faure et al., 2013). For instance, sterically stabilized silver nanoparticles capped by polyvinylpyrrolidone (PVP) do not agglomerate while electrostatically stabilized silver nanoparticles by citrate do agglomerate at low pH or high ionic strength (El Badawy et al., 2010). Electrosterically stabilized silver nanoparticles by branched polyethyleneimine capping are more resistant to agglomeration at low pH or high ionic strength as compared to citrate stabilized silver nanoparticles (El Badawy et al., 2010). Proteins such as bovine serum albumin (BSA), which stabilize silica nanoparticles through electrosteric mechanisms (Paula et al., 2014), are also known to stabilize chemically produced selenium nanoparticles (CheSeNPs) (Zhang et al., 2001). Proteins are also known to be associated with the BioSeNPs (Dobias et al., 2011; Lenz et al.,

2011). It has been proposed that BioSeNPs are coated with an organic layer of microbial origin, composed not exclusively of proteins (Winkel et al., 2012). However, to the best of our knowledge, the origin of this organic layer and its effect on the surface charge and thus, on the colloidal properties of BioSeNPs, is not known.

This study hypothesized that extracellular polymeric substances (EPS) are the capping agents and thus can affect the surface charge of the BioSeNPs that are available extracellularly. EPS are high molecular weight macromolecules that contain mainly proteins, carbohydrates, humic-like substances and small concentrations of DNA (Sheng et al., 2010; More et al., 2014). Thus, they provide many sites that can interact with the elemental selenium. EPS are an important component of mixed microbial aggregates i.e. biofilms or anaerobic granules employed for the treatment of selenium rich wastewaters in bioreactors (Flemming and Wingender, 2010; Sheng et al., 2010; Dhanjal and Cameotra, 2011). Besides, reduction of selenite to BioSeNPs by pure cultures has been reported both in the periplasmic space (Li et al., 2014) and extracellularly (Jiang et al., 2012), which further indicates that the BioSeNPs are likely to be grown in the presence of EPS.

In this study, the presence of an organic layer on the surface of BioSeNPs was determined from Energy Disperse X-ray spectroscopy (EDXS) and ζ-potential measurements as well as acid-base titrations. The presence of proteins and carbohydrates, suggesting the presence of EPS on the BioSeNPs, was confirmed by Fourier-transform Infrared Spectroscopy (FT-IR) and colorimetric measurements. EPS extracted from selenite fed anaerobic granules was used as a capping agent for CheSeNPs (EPS capped CheSeNPs) and their colloidal properties were studied as well. BSA, a well-known capping agent for CheSeNPs (Tran and Webster, 2011; Zhang et al., 2001), was used as a reference material to (a) demonstrate the capping ability of EPS and (b) show that the capping of BioSeNPs does not exclusively consists of proteins.

3.2. Materials and methods

3.2.1. BioSeNPs production and purification

BioSeNPs were produced using an anaerobic granular sludge treating pulp and paper wastewater, which has been described in detail by Roest et al. (2005). Anaerobic granular sludge (13 g L^{-1} wet weight) was added to the oxygen-free growth medium (NH$_4$Cl 5.6, CaCl$_2$·2H$_2$O 0.1, KH$_2$PO$_4$ 1.8, Na$_2$HPO$_4$ 2.0, KCl 3.3, in mM) with 20.0 mM of sodium lactate and 5.0 mM of sodium selenite. The incubation was carried out at 30 °C and pH 7.3 for 14 days. The production of elemental selenium was confirmed by the appearance of a red color (Figure S2b in Appendix 1). The produced BioSeNPs were purified following the protocol developed in Dobias et al. (2011) with minor modifications. Briefly, the supernatant was decanted, followed by simple centrifugation (Hermle Z36HK) at 3,000 g and 4 °C for 15 minutes to separate the suspended biomass. The collected BioSeNPs present in the supernatant from the previous centrifugation step were concentrated by centrifugation (Hermle Z36HK) at 37,000 g and 4 °C for 15 minutes. The pellet was re-suspended in Milli-Q (18MΩ cm) water and purified by sonication (15 minutes at 23 KHz, Soniprep 150, UK) followed by hexane separation. The concentration of BioSeNPs was determined by dissolving them in concentrated HNO$_3$ and then measuring the Se concentration by ICP-MS (Jain et al., 2015).

3.2.2. Analysis of the chemical composition of the BioSeNPs' surface

30 mL of purified BioSeNPs (390 mg L^{-1}) were sonicated for 15 minutes at 23 KHz using Soniprep 150 (MSE, UK) sonicator. After the sonication, BioSeNPs were centrifuged at 37,000 g for 30 minutes at 4 °C (Hermle Z36HK). The supernatant was collected and analyzed for carbohydrates (phenol-sulfuric acid method)(DuBois et al., 1956), proteins and humic-like substances (modified lowry method by Fr¢lund et al., 1995). The DNA concentration in the supernatant was measured after precipitation of the DNA with iso-propanol and then measuring the absorbance of the pellet using a spectrophotometer at 260 nm.(Wu and Xi, 2009)

3.2.3 EPS extraction and characterization

EPS was extracted from anaerobic granules, which were fed with selenite and lactate, and incubated at 30 °C for 14 days, using the NaOH extraction method (Liu and Fang, 2002). It is important to note that the NaOH extraction method may slightly alter the molecular structure of the EPS (Sheng et al., 2010), which is, however, unavoidable. The total organic carbon (TOC) and total organic nitrogen (TN) content of the extracted EPS was determined using a total organic carbon analyzer (Shimadzu TOC-VCPN analyzer, Kyoto, Japan). 3D excitation (220 - 400 nm) and emission (300 - 500 nm) fluorescent spectroscopy of extracted EPS (total organic carbon concentration 0.5 mg L^{-1}) was carried out using a FluoroMax-3 spectrofluorometer (HORIBA Jobin Yvon, Edison, NJ, USA) instrument. The carbohydrate, protein, humic-like substances and DNA content in the EPS were determined as described above. The FT-IR spectra of EPS were recorded on a Bruker Vertex 70/v spectrometer equipped with a D-LaTGS-detector (L-alanine doped triglycine sulfate) (more details in Appendix 1).

3.2.4. Production and purification of CheSeNPs, EPS capped CheSeNPs and BSA capped CheSeNPs

CheSeNPs were produced by reduction of sodium selenite (100 mM, 0.35 mL) by L-reduced glutathione (GSH) (100 mM, 1.4 mL) in a total volume of 30 mL at 22 $^{\circ}$C. EPS capped CheSeNPs and BSA capped CheSeNPs (chemically produced selenium nanoparticles in the presence of EPS and BSA, respectively) were produced in a similar manner, but in the presence of 100 mg L^{-1} total organic carbon of the extracted EPS and 100 mg L^{-1} of BSA, respectively. After the addition of NaOH (1 M) to adjust the pH to 7.2, the produced CheSeNPs, EPS capped CheSeNPs and BSA capped CheSeNPs were dialyzed against Milli-Q water (18MΩ cm) using a 3.5 kDa regenerated cellulose membrane while changing water every 12 hours for 96 hours (Zhang et al., 2001).

3.2.5. Selenium nanoparticles characterization

BioSeNPs were characterized by scanning electron microscopy (SEM) coupled with EDXS, ζ-potential measurements, hydrodynamic diameter (HDD) measurements, FT-IR and acid-base titrations (see Appendix 1 for more details). CheSeNPs were characterized by transmission electron microscopy (TEM) coupled with EDXS (TEM-EDXS) (see Appendix 1 for more details). BSA and EPS coated CheSeNPs were characterized by TEM-EDXS, ζ-potential measurements, HDD measurements and

FT-IR spectra (see Appendix 1 for more details). EPS and BSA capped CheSeNPs were contacted with different initial concentrations of Zn for ζ-potential measurements (see Appendix 1 for more details).

3.3. Results

3.3.1. SEM-EDXS analysis of BioSeNPs

The red BioSeNPs synthesized by the reduction of SeO_3^{2-} by anaerobic granular sludge are primarily spherical in shape (Figure S2a in Appendix 1). The resultant BioSeNPs formed a stable colloidal suspension (Figure S2b in Appendix 1). EDXS analysis of the BioSeNPs confirmed the presence of selenium (Figure S2c in Appendix 1). In addition, carbon, nitrogen, oxygen as well as weak signals of phosphorous, sulfur, calcium, and iron were also observed. The presence of carbon, nitrogen, oxygen, phosphorous and sulfur may be attributed to the EPS coating of the BioSeNPs, while calcium and iron can be likely traced back to the anaerobic granules used for the production of BioSeNPs.

3.3.2. Determination of functional groups present on the surface of BioSeNPs

The carbohydrates, proteins, humic-like substances and DNA concentrations released in the supernatant after the sonication of purified BioSeNPs were, respectively, 313.8 ± 3.5, 144.1 ± 2.1, 158.2 ± 2.3 and 4.6 ± 0.8 mg g^{-1} of BioSeNPs, confirming the presence of EPS components (Sheng et al., 2010) on the surface of the BioSeNPs. Acid-base titrations were carried out to determine the pK$_a$ values of the different functional groups present on the surface of the BioSeNPs. The acid-base titration curves for BioSeNPs (Figure S3 in Appendix 1) showed smoother lowering of the pH, as compared to the control, with the addition of acid (HCl). This can be attributed to the buffering capacity of the BioSeNPs due to the presence of various functional groups on their surface (Wang et al., 2012).

To evaluate the buffering capacity of BioSeNPs in more detail, the derivative of the acid-base titration was plotted with pH (Figure 3.1). The local minima represent the minimum variation of the pH and hence the buffering zones due to the adsorption of

H$^+$ ions on the surface of BioSeNPs, which corresponds to pK$_a$ values of the functional groups present on the surface of the BioSeNPs (Braissant et al., 2007). It is important to note that local minima will not be observed prominently in the beginning and the end of titration due to the relatively small pH change. The different pK$_a$ values and their corresponding functional groups are detailed in Figure 3.1. The presence of carboxylic acid (pK$_a$ = 3.9), phosphoric groups (pK$_a$ = 6.3) and sulfonic, sulfinic or thiol groups (pK$_a$ = 7.5) were confirmed by the acid-base titration (Figure 3.1) (Wang et al., 2012). Most notable absents are the amino groups, but this may be due to a small change in the pH value at the beginning of the titrations.

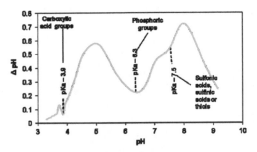

Figure 3.1. Derivative of acid-base titration data of BioSeNPs (—).

3.3.3. Characterization of EPS, EPS capped CheSeNPs, BSA capped CheSeNPs and CheSeNPs

The total organic carbon and total nitrogen concentration of the extracted EPS was 116.7 ± 0.5 mg L^{-1} and 19.0 ± 0.3 mg L^{-1}, respectively. 3D excitation emission fluorescent spectroscopy of EPS confirmed the presence of aromatic proteins and humic substances (Figure S4 in Appendix 1). The carbohydrates, proteins, humic-like substances and DNA concentrations in the extracted EPS were, respectively, 106.9 ± 2.3, 239.5 ± 6.2, 184.7 ± 15.3 and 2.7 ± 0.6 mg L^{-1}.

The extracted EPS and BSA were used as a capping agent for CheSeNPs produced by the reduction of sodium selenite using reduced-glutathione (Zhang et al., 2001). The addition of EPS during the chemical synthesis of selenium nanoparticles led to the formation of a clear suspended yellow-red colloidal solution (Figure 3.2a). A comparable result was obtained for BSA stabilized CheSeNPs (Figure 3.2b). In contrast, selenium nanoparticles produced in the absence of EPS formed a turbid brownish suspension (Figure 3.2c) in an hour, but settled only within two days. No agglomeration and settling of EPS and BSA capped CheSeNPs were observed even two weeks after their formation. It is important to note that the selenium concentration in the BSA capped CheSeNPs, EPS capped CheSeNPs and CheSeNPs was identical (1.1 mM). TEM micrographs revealed that EPS and BSA capped CheSeNPs were spherical in shape (Figure 3.2d, e). In contrast, wires of selenium were formed in the absence of EPS (Figure 3.2f). Obviously, EPS acted as a capping agent, thereby influencing the morphology of the CheSeNPs (Shah et al., 2010; Zhang et al., 2010; Zheng et al., 2012). EDX spectra confirmed the presence of selenium in all three types of CheSeNPs investigated (Figure 3.2g, h, i).

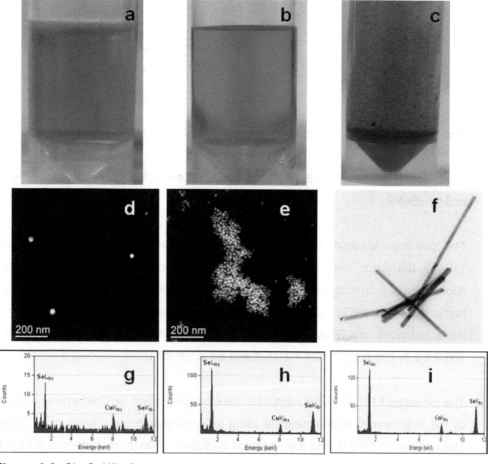

Figure 3.2. CheSeNPs formed in the presence of the capping agents (a) EPS and (b) BSA as well as (c) in the absence of any capping agent. High-angle annular dark-

field-scanning TEM micrographs of (d) EPS and (e) BSA stabilized CheSeNPs and (f) bright-field TEM micrograph of unstabilized CheSeNPs (no capping agent). EDX spectra obtained from (g) a single spherical EPS capped CheSeNp, (h) an ensemble of spherical BSA capped CheSeNPs and (i) from CheSeNPs nanowires prepared in the absence of EPS or BSA. Note that the Cu signal in the EDX spectra is caused by the carbon-coated copper support grid used for the TEM analysis.

3.3.4. Comparison of the capping agents on BioSeNPs, EPS capped CheSeNPs and BSA capped CheSeNPs

3.3.4.1. FT-IR analysis

FT-IR spectroscopy provided information of the functional groups present on the BioSeNPs (Figure 3.3, Table S1 in Appendix 1). The BioSeNPs had a broad feature between 3404 to 3270 cm^{-1}, corresponding to -OH and -NH stretching vibrations of amine and carboxylic groups (Xu et al., 2011). Additionally, small but sharp features at 2959, 2928 and 2866 cm^{-1} are observed which can be attributed to aliphatic saturated C-H stretching modes (Wang et al., 2012). The strongest feature is observed at 1646 cm^{-1} and mainly represents the stretching vibration of C=O present in proteins (amide I) (Xu et al., 2011). Thus, the feature at 1542 cm^{-1} corresponds to the N-H bending vibration in amide linkage of proteins (amide II) (Wang et al., 2012). The band around 1460 cm^{-1} might correspond to methyl groups and/or carboxylate groups (antisymmetric stretching vibration), whereas the feature at 1394 cm^{-1} is most likely attributed to the symmetric stretching vibration (Zhu et al., 2012). The presence of carboxylic groups is also evidenced by the weak shoulder observed around 1720 cm^{-1}. C-N stretching and N-H bending vibrations might contribute to the band observed at 1242 cm^{-1} (amide-III) (Wang et al., 2012). The small band at 1151 cm^{-1} is characteristic for P=O stretching modes (Wang et al., 2012). Another broad feature was observed between 1073 and 1038 cm^{-1}, which corresponds to C-O-C and C-H stretching arising from the carbohydrate groups (Wang et al., 2012; Zhu et al., 2012). The presence of S-H or S-O groups cannot be fully ascertained in the IR spectra due to their weak intensities or due to overlapping bands, respectively.

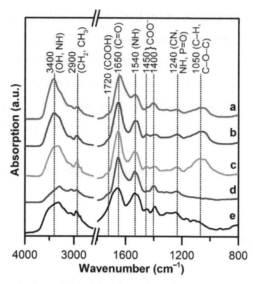

Figure 3.3. IR spectra of (a) EPS stabilized CheSeNPs, (b) extracted EPS, (c) BioSeNPs, (d) BSA stabilized CheSeNPs and (e) BSA. Indicated values are in cm^{-1}.

FT-IR spectra for BioSeNPs, EPS and EPS capped CheSeNPs were similar (Figure 3.3) and confirmed the presence of proteins (amide I: 1653-1646 cm^{-1}, amide II: 1537-1542 cm^{-1} and amide III: 1236-1242 cm^{-1}) and carbohydrates (1040-1077 cm^{-1}) (Figure 3.3 a-c). In contrast, the absence of carbohydrates for BSA and BSA stabilized CheSeNPs is shown in the IR spectra by the missing bands in the lower frequency range (Figure 3.3 d,e). The absence of carbohydrates residues is obviously reflected in the OH-stretching region (~3400 cm^{-1}) where reduced band intensities are observed in these spectra (Wang et al., 2012).

The overall shape of the FT-IR spectra of EPS capped CheSeNPs and EPS was very similar (Figure 3.3 a,b). Small spectral deviations were observed at 3400 cm^{-1} for EPS which were shifted to 3420 cm^{-1} for EPS capped CheSeNPs. The relative intensities of the features around 1450 and 1400 cm^{-1} are more different for EPS capped CheSeNPs than for EPS. This indicates a change of the carboxylate functional groups due to the presence of the selenium nanaoparticles. Interestingly, the same spectral feature was also observed for BSA stabilized CheSeNPs and BSA (Figure 3.3 d,e).

3.3.4.2. ζ-potential and HDD measurements

The ζ-potential of the BioSeNPs was −41.6 ± 0.5 mV at pH 7.0 and 1 mM NaCl concentration. At the same pH, but with 10 and 100 mM NaCl, the ζ-potential of the BioSeNPs was −29.7 ± 0.7 mV and −17.5 ± 0.9 mV, respectively. Similar negative ζ-potential values were reported for BioSeNPs formed at ambient temperature by bacterial cultures, i.e. *Bacillus selenatarsenatis* and *Bacillus cereus* (Dhanjal and Cameotra, 2010; Buchs et al., 2013). At a 10 and 100 mM background NaCl concentration, the iso-electric point of studied BioSeNPs was, respectively, 3.2 ± 0.1 mV and 2.4 ± 0.1 mV (Figure 3.4a, b).

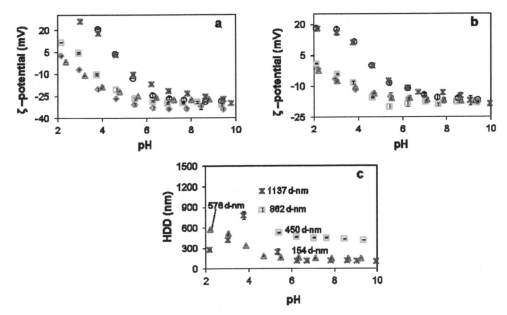

Figure 3.4. ζ-potential measurement of BioSeNP (□), EPS (◊), EPS capped CheSeNPs (△), BSA (○) and BSA capped CheSeNPs (∗) versus pH at (a) 10 mM and (b) 100 mM NaCl background electrolyte concentrations. (c) Hydrodynamic diameter of BioSeNP (□), EPS capped CheSeNPs (△) and BSA capped CheSeNPs (∗) versus pH at 100 mM NaCl background electrolyte concentration.

The ζ-potential of EPS, EPS capped CheSeNPs, BSA and BSA capped CheSeNPs was, respectively, −35 ± 1.1 mV, −31.5 ± 0.8 mV, −42.9 ± 3.2 mV and −38 ± 0.4 mV at 1 mM NaCl concentration and neutral pH. The ζ-potential versus pH curve showed an iso-electric point at pH 2.3 ± 0.1 mV for both EPS and EPS capped CheSeNPs at 10 and 100 mM NaCl background electrolyte concentrations (Figure 3.4a, b). It was not possible to determine the ζ-potential versus pH curve of uncapped CheSeNPs as a small contamination of glutathione affected the ζ-potential versus pH profile. In this experiment, the profile of BSA and BSA capped CheSeNPs are a positive control as BSA is known to stabilize the selenium nanoparticles (Zhang et al., 2001; Tran and Webster, 2011) with a reported iso-electric point of 4.6 ± 0.1 mV (Salg et al., 2012) (Figure 3.4a, b).

The ζ-potential versus pH profiles of BioSeNPs, EPS, EPS capped CheSeNPs, BSA and BSA capped CheSeNPs are similar from pH 9.5 to 6.0. However, at pH values below 6.0, BioSeNPs, EPS and EPS capped CheSeNPs follow similar profiles and remain more negative than BSA and BSA capped CheSeNPs (Figure 3.4). This leads to a similar iso-electric point of BioSeNPs, EPS and EPS capped CheSeNPs (~pH_{IEP} 2.4 at 100 mM NaCl concentration) as compared to 4.6 observed for BSA and BSA capped CheSeNPs. It is important to note that the iso-electric point of BioSeNPs (pH_{IEP} 3.2 ± 0.1) at 10 mM NaCl background electrolyte was slightly different than the iso-electric point (pH_{IEP} 2.4 ± 0.1) of BioSeNPs at 100 mM NaCl background electrolyte. This change is not significant enough to unambiguously conclude that the differences are due to the interfering background electrolyte.

It has been shown that the ζ-potential of BioSeNPs loaded with Zn becomes less negative leading to a lowering of their colloidal stability (Jain et al., 2015). Similar experiments with BSA and EPS capped CheSeNPs suggested that loading of Zn on these CheSeNPs also lead to a less negative ζ-potential of −7.3 and −11.0 mV for, respectively, BSA and EPS capped CheSeNPs contacted with 1000 mg L^{-1} Zn (Figure S5 in Appendix 1). The equilibrium pH varied between 5.5 and 6.5. The zinc concentration required to achieve −5 to −10 mV for BSA and EPS capped CheSeNPs was 10 times more than that required for BioSeNPs, even when the concentration of selenium concentration was 4 times higher in the BioSeNPs (Jain et al., 2015). This might be due the smaller size of BSA and EPS capped CheSeNPs (30-50 d-nm) as compared to BioSeNPs (180 d-nm) (Jain et al., 2015).

The HDD of BioSeNPs increased slightly from 403 ± 8 to 531 ± 6 d-nm when the pH changed from 10.2 to 5.8 at 100 mM NaCl concentration. However, when the pH dropped to 4.7 and 3.8, the HDD increased to 862 ± 29 and 2130 ± 180 d-nm, respectively (Figure 3.4c). The HDD of EPS capped CheSeNPs and BSA capped CheSeNPs also increased as the pH decreased (Figure 3.4c). There is a large jump in the HDD of BSA capped CheSeNPs as the pH approached the iso-electric point of BSA. A similar, but smaller jump, in HDD of EPS capped CheSeNPs is observed as the pH approaches the iso-electric point of EPS. HDD measurements at 10 mM NaCl background electrolyte (Figure S6 in Appendix 1) gave similar profiles as observed in Figure 3.4c. It is important to note that it was not possible to compare the HDD versus pH profile of CheSeNPs free of capping agents, as these CheSeNPs had a different shape (wire versus sphere) and they sediment as compared to the capped CheSeNPs (Figure 3.2).

3.4. Discussion

3.4.1. EPS are present on the BioSeNPs

This study suggested, for the first time, that the organic layer present on the BioSeNPs synthesized by anaerobic granules is the EPS. The presence of the various functional groups on the BioSeNPs' surface (Figure 3.1, 3) is due to the attached organic polymers, most likely produced by the microorganisms present in the anaerobic granules. The presence of carbohydrates, proteins and humic-like substances, which were released upon sonication of the purified BioSeNPs, suggests that EPS comprising of these components is present on the surface of the BioSeNPs. The presence of EPS was further suggested by the presence of amide-I and amide-II bands (proteins) and strong bands at 1073 and 1038 cm^{-1} (carbohydrates) in the IR spectra (Figure 3.3, Table S1 in Appendix 1). The presence of these carbohydrates, which are always a part of EPS and are not observed in IR spectra of pure proteins (Kong and Yu, 2007; Wang et al., 2012), indicates that not protein but EPS containing both proteins and carbohydrates, are present on the BioSeNPs (Figure 3.3, Table S1 in Appendix 1). Moreover, the similar overall shape of the IR spectra and ζ-potential versus pH variation of BioSeNPs, EPS and EPS capped CheSeNPs (Figures 3 and 4a, b) further confirms the presence of EPS on the surface of BioSeNPs. The small DNA concentration found on the surface of BioSeNPs rules out the possibility of cell lysis, thus further confirming the layer on BioSeNPs is from EPS and not due to intracellular organics from the microbial cells. This is an interesting finding as previously only the presence of proteins has been reported on the surface of BioSeNPs (Dobias et al., 2011; Lenz et al., 2011) and the origin of the organic layer on the BioSeNPs was unknown(Winkel et al., 2012).

The similar overall shape of the amide-I and –II modes in IR spectra of EPS and EPS capped CheSeNPs suggests (Figure 3.3) that the EPS was attached to elemental selenium without major modification in its secondary structure. The distinct shift of some spectral features, e.g. shifting of features from 3400 to 3420 cm^{-1} and 1384 to 1405 cm^{-1} in EPS capped CheSeNPs as compared to EPS cannot be unambiguously attributed to the interaction of hydroxyl or carboxylic acid groups(Zhu et al., 2012) with elemental selenium, respectively, although such assignments appear to be obvious. For instance, a slightly different water content in the EPS capped CheSeNPs and EPS samples might cause similar shifts in the spectrum of a KBr pellet used in the sample preparation for FT-IR analysis. However, the band at 1452 cm^{-1} observed in the spectrum of EPS, but not in the EPS capped CheSeNPs spectrum, most likely represents the antisymmetric stretching mode of carboxylate groups. Thus, the disappearance of this mode reflects the interaction of elemental selenium with these functional groups of EPS. The interaction of EPS and elemental selenium with hydroxyl groups suggests the likely interaction of the carbohydrate fraction of the EPS with elemental selenium, as the sugar residues can be expected to show much more -OH groups. However, further research is required to identify the exact EPS fractions interacting with the CheSeNPs and BioSeNPs.

The synthesis of BioSeNPs involves two steps: (1) the reduction of selenite to elemental selenium and (2) the subsequent growth of elemental selenium to 50-250 d-nm BioSeNPs with a median of 180 d-nm (Jain et al., 2015) (Figure 3.5). Reduction of selenite to elemental selenium is carried out intracellularly (Debieux et al., 2011), in the periplasmic space (Li et al., 2014) or extracellulary (Jiang et al., 2012) in different microorganisms and thus, the growth of BioSeNPs can take place either intracellularly or extracellularly. It should be noted that this study did not distinguish between the BioSeNPs that grew intracellularly, which are then subsequently expelled outside in the supernatant from those which grew extracellularly. For the elemental selenium that was formed extracellularly, the growth to BioSeNPs takes place in the presence of EPS which is per definition present extracellularly. The elemental selenium, which was formed intracellularly, will most likely have a coating of predominantly proteins that are involved in their production and subsequent secretion (Debieux et al., 2011). These intracellular elemental selenium particles may partially grow to BioSeNPs coated with proteins (Debieux et al., 2011). The size of these BioSeNPs secreted from the cell is likely to be much smaller (considering the BioSeNPs transport through the cell without damaging the membrane or cell wall). Since the average size of BioSeNPs observed in the majority of the studies is greater than 150 d-nm (Winkel et al., 2012; Jain et al., 2014), further growth of the excreted BioSeNPs can then take place extracellularly, in the presence of the EPS. For the BioSeNPs produced and growing intracellularly but expelled outside the cell due to cell lysis, the full growth would have taken place in absence of EPS and the BioSeNPs are thus capped by predominantly proteins. The smaller quantity of DNA on the BioSeNPs suggests that cell lysis in this study was minimal, thus, the majority of the BioSeNPs observed in this study were formed or growing extracellularly in the presence of EPS.

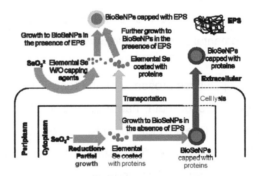

Figure 3.5. Scheme demonstrating the growth of elemental selenium to BioSeNPs.

3.4.2 .EPS govern the surface charge of BioSeNPs

The ζ-potential versus pH curves and iso-electric points of BioSeNPs, EPS and EPS capped CheSeNPs (Figure 3.4a, b) were similar, while also those of BSA and BSA capped CheSeNPs were identical (Figure 3.4a, b). This shows that the ζ-potential and iso-electric point of BioSeNPs and EPS capped CheSeNPs are governed by the surface charge of the EPS, rather than that of the elemental selenium. The almost identical iso-electric points of BioSeNPs, EPS and EPS capped CheSeNPs (Figure 3.4b) further suggests that the EPS is capping the BioSeNPs and CheSeNPs for the entire pH range tested.

Capping agents can provide the colloidal stability to nanoparticles either by electrostatic, steric or electrosteric mechanisms (El Badawy et al., 2010). The sudden jump in HDD of BioSeNPs and EPS capped CheSeNPs (Figure 3.4c and Figure S6 in Appendix 1) at a pH close to the iso-electric point indicates agglomeration of nanoparticles which is a hint for electrostatic stabilization by EPS. The dependence of the ζ-potential on the ionic strength further suggests that the EPS stabilize the BioSeNPs and CheSeNPs by electrostatic repulsion (Figure 3.4a and b) (El Badawy et al., 2010). However, due to the bulky structure of EPS, partial stabilization of EPS capped CheSeNPs by steric hindrance cannot be excluded, as demonstrated for the stabilization of silica nanoparticles and quantum dots by BSA (Bucking et al., 2010; Paula et al., 2014). Thus, this study strongly suggests that the EPS stabilizes the BioSeNPs and CheSeNPs both electrostatically and sterically, thus electrosterically. It is important to note that the larger HDD of BioSeNPs as compared to EPS capped CheSeNPs at a pH exceeding 6 might be due to suboptimum EPS to elemental selenium ratio during the formation of the BioSeNPs.

The carboxylic acid group, whose presence on the BioSeNPs' surface was confirmed by the IR and acid-base titration data, has a pK_a value of 3.9 (Figure 3.1, 3.3 and Table S1 in Appendix 1). Due to their low pK_a values, this group will be largely deprotonated at pH values above 5.5 (more than 98%, calculated using the Henderson-Hasselbalch equation). The negative ζ-potential value of the BioSeNPs at a pH below 5.5 suggests a large number of carboxylic acid group sites in the EPS capping the BioSeNPs (Figure 3.1 and 3.4).

3.4.3. Environmental implications

This study has demonstrated that EPS present in anaerobic granular sludge are capping the BioSeNPs, govern their surface charge and thus, affect their colloidal properties in the engineered settings where they are synthesized. It is important to point out that the intracellularly grown BioSeNPs will be capped by (a mixture of) proteins and thus their ζ-potential versus pH profile and iso-electric point might be different than that of the BioSeNPs growing extracellularly in the presence of EPS (Figure 3.4a, b and 3.5). This would lead to a different colloidal behavior of BioSeNPs that are capped with proteins and those that are capped with EPS in various environmental conditions (pH and interactions with heavy metals, Figure S5 in Appendix 1), thus, affecting their fate in the environment.

The presence of EPS on the surface of BioSeNPs makes them stable in the colloidal suspension and thus, mobile in the low ionic strength and neutral pH environment. This is in contrast to our understanding that EPS of biofilms restrict the dispersion of natural and engineered nanoparticles, as shown for Se (Bajaj et al., 2012), CdSe, Ag or ZnS nanoparticles (Tourney and Ngwenya, 2014). Thus, this study highlights the importance of further studies on the role of EPS in the fate of bioreduced products of redox active elements in both natural and engineered settings.

Acknowledgments

The authors are thankful to Dr. Graciella Gil-Gonzalez (KAUST, Saudi Arabia) for insightful discussion, Ferdi Battles (UNESCO-IHE, The Netherlands) for the Nano-sizer and acid-base titration experiments, Berend Lolkema (UNESCO-IHE, The Netherlands) for TOC, TN and 3D EEM analysis and Elfi Christalle (Helmholtz-Zentrum Dresden-Rossendorf, Germany) for SEM-EDXS measurements and Purvi Jain (Utrecht University, The Netherlands) for DNA measurements.

3.5 References

Bajaj, M., Schmidt, S., Winter, J., 2012. Formation of Se (0) nanoparticles by Duganella sp. and Agrobacterium sp. isolated from Se-laden soil of North-East Punjab, India. Microb. Cell Fact. 11, 64.

Braissant, O., Decho, a. W., Dupraz, C., Glunk, C., Przekop, K.M., Visscher, P.T., 2007. Exopolymeric substances of sulfate-reducing bacteria: Interactions with calcium at alkaline pH and implication for formation of carbonate minerals. Geobiology 5, 401–411.

Buchs, B., Evangelou, M.W.-H., Winkel, L., Lenz, M., 2013. Colloidal properties of nanoparticular biogenic selenium govern environmental fate and bioremediation effectiveness. Environ. Sci. Technol. 47, 2401–2407.

Bucking, W., Massadeh, S., Merkulov, A., Xu, S., Nann, T., 2010. Electrophoretic properties of BSA-coated quantum dots. Anal. Bioanal. Chem. 396, 1087–1094.

Cantafio, A.W., Hagen, K.D., Lewis, G.E., Bledsoe, T.L., Nunan, K.M., Macy, J.M., 1996. Pilot-scale selenium bioremediation of san joaquin drainage water with Thauera selenatis. Appl. Environ. Microbiol. 62, 3298–303.

Debieux, C.M., Dridge, E.J., Mueller, C.M., Splatt, P., Paszkiewicz, K., Knight, I., Florance, H., Love, J., Titball, R.W., Lewis, R.J., Richardson, D.J., Butler, C.S., 2011. A bacterial process for selenium nanosphere assembly. PNAS 108, 13480–13485.

Dhanjal, S., Cameotra, S.S., 2011. Selenite Stress Elicits Physiological Adaptations in Bacillus sp . (Strain JS-2). J. Microbiol. Biotechnol. 21, 1184–1192.

Dhanjal, S., Cameotra, S.S., 2010. Aerobic biogenesis of selenium nanospheres by Bacillus cereus isolated from coalmine soil. Microb. Cell Fact. 9, 52. Dobias, J., Suvorova, E.I., Bernier-latmani, R., 2011. Role of proteins in controlling selenium nanoparticle size. Nanotechnology 22, 195605. doi:10.1088/0957-4484/22/19/195605

Dobias, J; Suvorova, E.I, Bernier-latmani, R., 2011. Role of proteins in controlling selenium nanoparticle size. Nanotechnology 22, 195605.

DuBois, M., Gilles, K. a., Hamilton, J.K., Rebers, P. a., Smith, F., 1956. Colorimetric Method for Determination of Sugars and Related Substances. Anal. Chem. 28, 350–356.

El Badawy, A.M., Luxton, T.P., Silva, R.G., Scheckel, K.G., Suidan, M.T., Tolaymat, T.M., 2010. Impact of environmental conditions (pH, ionic strength, and electrolyte type) on the surface charge and aggregation of silver nanoparticles suspensions. Environ. Sci. Technol. 44, 1260–6.

Faure, B., Salazar-Alvarez, G., Ahniyaz, A., Villaluenga, I., Berriozabal, G., De Miguel, Y.R., Bergström, L., 2013. Dispersion and surface functionalization of oxide nanoparticles for transparent photocatalytic and UV-protecting coatings and sunscreens. Sci. Technol. Adv. Mater. 14, 023001.

Flemming, H.-C., Wingender, J., 2010. The biofilm matrix. Nat. Rev. Microbiol. 8, 623–33.

Frølund, B., Griebe, T., Nielsen, P.H., 1995. Enzymatic activity in the activated-sludge floc matrix. Appl. Microbiol. Biotechnol. 43, 755–761.

Hamilton, S.J., 2004. Review of selenium toxicity in the aquatic food chain. Sci. Total Environ. 326, 1–31. doi:10.1016/j.scitotenv.2004.01.019

Jain, R., Gonzalez-Gil, G., Singh, V., van Hullebuchs, Eric, D., Farges, F., Lens, P.N.L., 2014. Biogenic selenium nanoparticles : production , characterization and challenges, in: Kumar, A., Govil, J, N. (Eds.), Nanobiotechnology. Studium Press LLC, USA, pp. 361–390.

Jain, R., Jordan, N., Schild, D., Hullebusch, E.D. Van, Weiss, S., Franzen, C., Hubner, R., Farges, F., Lens, P.N.L., 2015. Adsorption of zinc by biogenic elemental selenium nanoparticles. Chem. Eng. J. 260, 850–863.

Jiang, S., Ho, C.T., Lee, J.-H., Duong, H. Van, Han, S., Hur, H.-G., 2012. Mercury capture into biogenic amorphous selenium nanospheres produced by mercury resistant Shewanella putrefaciens 200. Chemosphere 87, 621–4.

Kong, J., Yu, S., 2007. Fourier Transform Infrared Spectroscopic Analysis of Protein Secondary Structures. Acta Biochim. Biophys. Sin. (Shanghai). 39, 549–559.

Lenz, M., Hullebusch, E.D. Van, Hommes, G., Corvini, P.F.X., Lens, P.N.L., 2008. Selenate removal in methanogenic and sulfate-reducing upflow anaerobic sludge bed reactors. Water Res. 42, 2184–2194.

Lenz, M., Kolvenbach, B., Gygax, B., Moes, S., Corvini, P.F.X., 2011. Shedding light on selenium biomineralization: proteins associated with bionanominerals. Appl. Environ. Microbiol. 77, 4676–80.

Lenz, M., Lens, P.N.L., 2009. The essential toxin: the changing perception of selenium in environmental sciences. Sci. Total Environ. 407, 3620–33.

Li, D.-B., Cheng, Y.-Y., Wu, C., Li, W.-W., Li, N., Yang, Z.-C., Tong, Z.-H., Yu, H.-Q., 2014. Selenite reduction by Shewanella oneidensis MR-1 is mediated by fumarate reductase in periplasm. Sci. Rep. 4, 3735.

Liu, H., Fang, H.H.P., 2002. Extraction of extracellular polymeric substances (EPS) of sludges. J. Biotechnol. 95, 249–56.

More, T.T., Yadav, J.S.S., Yan, S., Tyagi, R.D., Surampalli, R.Y., 2014. Extracellular polymeric substances of bacteria and their potential environmental applications. J. Environ. Manage. 144, 1–25.

Oremland, R.S., Herbel, M.J., Blum, J.S., Langley, S., Beveridge, T.J., Ajayan, P.M., Sutto, T., Ellis, A. V, Curran, S., 2004. Structural and spectral features of selenium nanospheres produced by Se-respiring bacteria. Appl. Environ. Microbiol. 70, 52–60.

Paula, A.J., Silveira, C.P., Ste, D., Filho, A.G.S., Romero, F. V, Fonseca, L.C., Tasic, L., Alves, O.L., Dura, N., 2014. Topography-driven bionano-interactions on colloidal silica nanoparticles. Appl. Mater. interfaces 6, 3437–3447.

Qin, H.-B., Zhu, J.-M., Liang, L., Wang, M.-S., Su, H., 2013. The bioavailability of selenium and risk assessment for human selenium poisoning in high-Se areas, China. Environ. Int. 52, 66–74.

Rayman, M.P., 2000. The importance of selenium to human health. Lancet 356, 233–41.

Roest, K., Heilig, H.G.H.J., Smidt, H., de Vos, W.M., Stams, A.J.M., Akkermans, A.D.L., 2005. Community analysis of a full-scale anaerobic bioreactor treating paper mill wastewater. Syst. Appl. Microbiol. 28, 175–85.

Salg, S., Salgı, U., Bahad, S., 2012. Zeta potentials and isoelectric points of biomolecules : the effects of ion types and ionic strengths. Int. J. Electrochem. Sci 7, 12404–12414.

Shah, C.P., Singh, K.K., Kumar, M., Bajaj, P.N., 2010. Vinyl monomers-induced synthesis of polyvinyl alcohol-stabilized selenium nanoparticles. Mater. Res. Bull. 45, 56–62.

Sheng, G.-P., Yu, H.-Q., Li, X.-Y., 2010. Extracellular polymeric substances (EPS) of microbial aggregates in biological wastewater treatment systems: a review. Biotechnol. Adv. 28, 882–94.

Tourney, J., Ngwenya, B.T., 2014. The role of bacterial extracellular polymeric substances in geomicrobiology. Chem. Geol. 386, 115–132.

Tran, P. a, Webster, T.J., 2011. Selenium nanoparticles inhibit *Staphylococcus aureus* growth. Int. J. Nanomedicine 6, 1553–8.

USEPA, August 1st, 2014, National Recommended Water Quality Criteria, http://water.epa.gov/scitech/swguidance/standards/criteria/current/index.cfm

Wang, L.-L., Wang, L.-F., Ren, X.-M., Ye, X.-D., Li, W.-W., Yuan, S.-J., Sun, M., Sheng, G.-P., Yu, H.-Q., Wang, X.-K., 2012. pH dependence of structure and surface properties of microbial EPS. Environ. Sci. Technol. 46, 737–44.

Winkel, L.H.E., Johnson, C.A., Lenz, M., Grundl, T., Leupin, O.X., Amini, M., Charlet, L., 2012. Environmental selenium research: from microscopic processes to global understanding. Environ. Sci. Technol. 46, 571–579.

Wu, J., Xi, C., 2009. Evaluation of different methods for extracting extracellular DNA from the biofilm matrix. Appl. Environ. Microbiol. 75, 5390–5.

Xu, C., Zhang, S., Chuang, C., Miller, E.J., Schwehr, K. a., Santschi, P.H., 2011. Chemical composition and relative hydrophobicity of microbial exopolymeric substances (EPS) isolated by anion exchange chromatography and their actinide-binding affinities. Mar. Chem. 126, 27–36.

Zhang, J.S., Gao, X.Y., Zhang, L.D., Bao, Y.P., 2001. Biological effects of a nano red elemental selenium. Biofactors 15, 27–38.

Zhang, Y., Wang, J., Zhang, L., 2010. Creation of highly stable selenium nanoparticles capped with hyperbranched polysaccharide in water. Langmuir 26, 17617–23.

Zheng, S., Li, X., Zhang, Y., Xie, Q., Wong, Y.-S., Zheng, W., Chen, T., 2012. PEG-nanolized ultrasmall selenium nanoparticles overcome drug resistance in hepatocellular carcinoma HepG2 cells through induction of mitochondria dysfunction. Int. J. Nanomedicine 7, 3939–49.

Zhu, L., Qi, H., Lv, M., Kong, Y., Yu, Y., Xu, X., 2012. Component analysis of extracellular polymeric substances (EPS) during aerobic sludge granulation using FTIR and 3D-EEM technologies. Bioresour. Technol. 124, 455–9.

CHAPTER 4

Biogenic synthesis of elemental selenium nanowires

This chapter will be published as

Jain, R., Jordan, N., Kacker, R., Weiss, S., Hübner, R., Kramer, H., van Hullebusch, E.D., Farges, F., Lens, P.N.L, 2014. Biogenic synthesis of elemental selenium nanowires *(in preparation)*.

Abstract:

The microbial reduction of selenium oxyanions always lead to the formation of amorphous or monoclinic colloidal elemental spherical selenium nanoparticles. This study demonstrated the production and characterization of biogenic trigonal elemental selenium nanowires (BioSeNWs). Selenite was reduced in the presence of anaerobic granules under anaerobic conditions at thermophilic conditions (55 and 65 °C). The produced BioSeNWs were purified using a protocol developed in an earlier study. Scanning electron microscopy (SEM) revealed the median diameter of 20-30 nm and 40-50 nm for BioSeNWs produced at 55 and 65 °C, respectively. Energy disperse X-ray specta confirmed that BioSeNWs were mainly composed of selenium. The produced BioSeNWs were in trigonal crystalline state as confirmed by raman spectroscopy and X-ray diffraction. The BioSeNWs were colloidally stable owning to negative ζ-potential. The negative ζ-potential of the BioSeNWs was due to the presence of organic layer most likely originating from the extracellular polymeric substances (EPS).

Keywords: Biogenic, nanowires, selenium, trigonal, EPS

Graphical abstract:

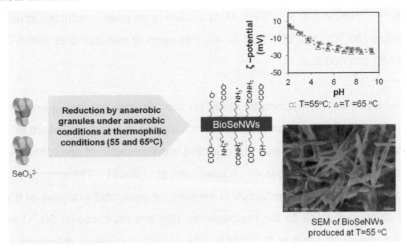

SEM of BioSeNWs
produced at T=55 °C

4.1. Introduction

Elemental selenium displays many unique properties such as high photoconductivity, catalytic activities as well as piezoelectric and thermoelectric effects (Gates et al., 2002) These properties become more pronounced and effective as the size of the particles decreases (< 100 nm) to form nanoparticles (Gates et al., 2002; Shah et al., 2010) Therefore, selenium nanoparticles are used in solar cells, semiconductor rectifiers, xerography and other functional materials (Gates et al., 2002) Of the different selenium nanoparticles that can be synthesized, the one dimensional nanowires are very attractive for material scientists as these nanowires can be used as connectors in the fabrication of nanodevices as well as to study the effects on the mechanical, optical and electrical properties due to the size confinement (Gates et al., 2002).

In general, biologically produced nanoparticles can be a good choice to replace chemically produced nanoparticles due to their relatively easy, non-toxic and green production approach and low cost (Faramarzi and Sadighi, 2013; Kharissova et al., 2013; Wang et al., 2012). The formation of spherical selenium nanoparticles has been reported widely as the microbial reduction of selenite or selenate always results in the formation of amorphous or monoclinic spherical selenium nanoparticles (Jain et al., 2014; Oremland et al., 2004). Most studies have been conducted at mesophilic temperature (30 $^{\circ}$C), and all report on the formation of nanospheres with a diameter between 50 and 400 nm.

There is, however, no systematic study on the morphology and characteristics of biogenic formed selenium nanoparticles. For instance, a simple exposure to elevated temperatures is an effective way to trigger the transformation of amorphous selenium to trigonal selenium nanowires (Gates et al., 2002). Therefore, this study investigated the anaerobic reduction of selenite by anaerobic granules at 55 and 65 $^{\circ}$C and compared these to the nanospheres that are produced at 30 $^{\circ}$C that were described previously (Jain et al., 2014a). The produced biogenic elemental selenium nanowires (BioSeNWs) were also compared for their shape, size, colloidal stability and crystallinity with chemically synthesized elemental selenium nanoparticles (CheSeNPs) using scanning electron microscopy - energy disperse X-ray spectroscopy (SEM-EDXS), X-ray diffraction and Raman spectroscopy. The surface and optical properties of the BioSeNWs were also characterized using zetametry, acid-base titration and FT-IR.

4.2. Materials and methods

4.2.1. Production and purification of BioSeNWs and CheSeNPs

BioSeNWs were produced by the reduction of selenite by anaerobic granular sludge using lactate as electron donor under anaerobic incubation at 55 and 65 °C for 7 days. The composition of the medium (pH 7) and the biomass is described in (Jain et al., 2015). BioSeNWs were purified as described in Jain et al. (2014a). Briefly, the formation of BioSeNWs was observed by the appearance of greyish color in the incubated bottles. The biomass was separated from the BioSeNWs by simple decanting followed by centrifugation at 3,000g and then concentrated by centrifugation at 37,000g. This step was followed by resuspension of the pellet in Milli-Q water (18 MΩ*cm), followed by sonication and hexane separation. The collected aqueous phase was washed thrice with Milli-Q water. The concentration of selenium was determined by Inductively Coupled Plasma-Mass Spectrometry (ICP-MS) as described in Jain et al. (2015). CheSeNPs were produced at ambient temperature by reduction of selenite by L-glutathione reduced in absence of any capping agent and then purified by dialysis against Milli-Q water using a 3.5 kDa regenerated cellulose membrane while changing water every 12 hours for 96 hours (Zhang et al., 2001).

4.2.2. Characterization of BioSeNWs

SEM-EDXS, FT-IR, acid-base titration and ζ-potential measurements were carried out as described in Jain et al. (2014a). XRD analysis was carried out as described in Jain et al. (2015). Raman measurements were performed at room temperature with a Bruker Vertex 70/v vacuum FT-IR spectrometer equipped with a Ge detector where a FT-Raman module (Nd-YAG laser, $\lambda_{exc.}$= 1064 nm, P = 450 mW) has been implemented. Spectra were averaged out of 256 scans for BioSeNWs produced at 55 and 65 °C and out of 32 scans for CheSeNPs.

4.3. Results

4.3.1. Shape, size and crystallinity of BioSeNWs

Anaerobic reduction of selenite to elemental selenium was observed by the appearance of grey color in the supernatant for both 55 and 65 °C . SEM images confirmed the formation of BioSeNWs with a median diameter of 20-30 nm and 40-50 nm at 55 and 65 °C, respectively (Figure 1a, b, g). The length to diameter ratio of the BioSeNWs exceeded 20. The median diameter of the CheSeNPs was 30-40 nm (Figure 1c, f) and their length to diameter ratio was less than that observed for BioSeNWs.

EDXS analysis revealed that BioSeNWs are mainly composed of elemental selenium. In addition, carbons, oxygen, sulfur and phosphorous were presentin the BioSeNWs produced at both 55 and 65 °C (Figure 1d, e). The presence of these elements can be attributed to presence of extracellular polymeric substances that are formed as a coating on the BioSeNWs. The presence of iron, as revealed by EDXS analysis, can be attributed its presence in the inoculum anaerobic granular sludge. EDXS analysis of CheSeNPs revealed the presence of only selenium, carbon and oxygen. The presence of carbon and oxygen is possibly due to the use of the carbon coating used for the preparation of the SEM samples.

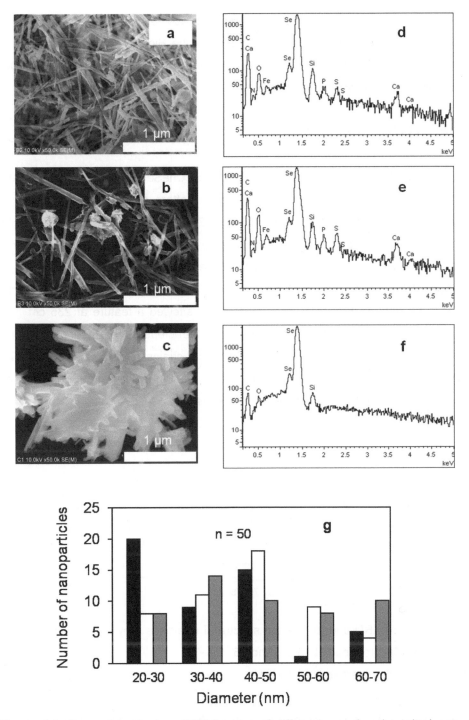

Figure 4.1. Secondary electron SEM images of different samples deposited onto a piece of Si wafer: BioSeNWs prepared at (a) 55 °C, (b) 65 °C and (c) CheSeNPs

produced at ambient temperature and their representative EDXS analysis (d), (e) and (f), respectively. (f) Diameter distribution of the BioSeNWs produced at 55 °C (■), 65 °C (□) and CheSeNPs (■). Note that the presence of silicon in the EDXS analysis is due the use of silicon wafer for holding of the samples.

4.3.2. Crystallinity of BioSeNWs

The BioSeNWs produced at 55 and 65 °C and CheSeNPs all had identical XRD patterns. All peaks observed in the diffraction pattern of both BioSeNWs can be indexed to trigonal phase of selenium (Figure 2a). The intensity of the peaks at the 2-theta values of 23.5 and 29.5 are comparable and all the other peaks are minor for the BioSeNWs. In contrast, the intensity of the peak at theta value 29.5 2- for the CheSeNPs is higher than the one observed at theta value 23.5 2. The Raman spectra of both the BioSeNWs and CheSeNPs showed a feature at 235 cm^{-1} and 140 cm^{-1} (Figure 2b). These features are characteristics of trigonal elemental selenium arising from the vibration of Se-helical chains (Gates et al., 2002). The XRD and Raman spectra thus confirmed that the BioSeNWs and CheSeNPs are largely present in form of trigonal elemental selenium.

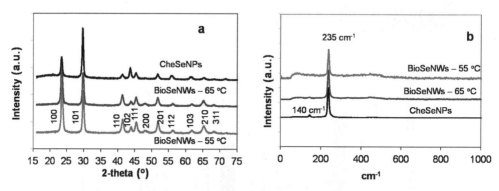

Figure 4.2. (a) XRD and (b) Raman spectra of reference grey trigonal selenium (–) and BioSeNWs produced at 55 (–) and 65 °C (–).

4.3.3. Colloidal properties of BioSeNWs

The produced BioSeNWs were present in the form of a colloidal suspension unlike the CheSeNPs at similar concentraton of 3.8 mM (Figure 3a, b, c). The ζ-potential of

BioSeNWS produced at 55 and 65 °C was −39 and −40 mV, respectively, at pH 7 and 1 mM background electrolyte concentration. The negative ζ-potential of the elemental selenium produced by *Bacillus selenatarsenatis*, *Bacillus cereus* as well as anerobic granules have been reported (Buchs et al., 2013; Dhanjal and Cameotra, 2010; Jain et al., 2015). However, all these studies reported the formation of nanospheres, unlike nanowires as reported in this study. The iso-electric point of BioSeNW produced at 55 and 65 °C was 2.7 and 2.6, respectively, at 10 mM background electrolyte concentration (Figure 3d). It is important to note that it is not possible to remove all the glutathione used for the formation of CheSeNPs which would have impacted the ζ-potential measurements of CheSeNPs and thus, the ζ-potential versus pH profile of the CheSeNPs is not reported.

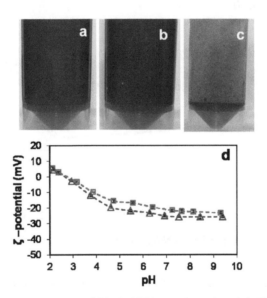

Figure 4.3. Colloidal suspension of BioSeNWs produced at (a) 55 °C and (b) 65 °C and (c) settled CheSeNPs. (d) ζ-potential versus pH profile of of BioSeNWs produced at 55 °C (□) and 65 °C (Δ).

4.3.4. Functional groups associated with the BioSeNWs

BioSeNWs produced at 55 and 65 °C exhibit the same features as BioSeNPs produced at 30 °C except a feature missing at 1394 cm^{-1} (Figure 4) (Jain et al., 2014a). Briefly, FT-IR confirmed the presence of OH (3404 to 3270 cm^{-1}), C$_x$H$_y$ (2959 - 2866 cm^{-1}), COO (1720 cm^{-1}), C=O (1646 cm^{-1}), N-H (1542 cm^{-1}), C-N

(1242 cm^{-1}), P=O (1151 cm^{-1}) and C-O-C and C-H (1073 - 1038 cm^{-1}) stretching or vibrations. The features at 1646 cm^{-1} and broad feature between 1073 - 1038 cm^{-1} suggests the presence of proteins and carbohydrates on the BioSeNWs. The absence of a feature at 1394 cm^{-1} when compared to BioSeNPs produced at 30 °C may be explained by the fact that the features at 1459 and 1398 cm^{-1} combined to give the feature at 1438 and 1449 cm^{-1} for BioSeNWs produced at 55 and 65 °C, respectively, corresponding to the methyl groups or the symmetric stretching vibration of deprotonated carboxylic acid groups.

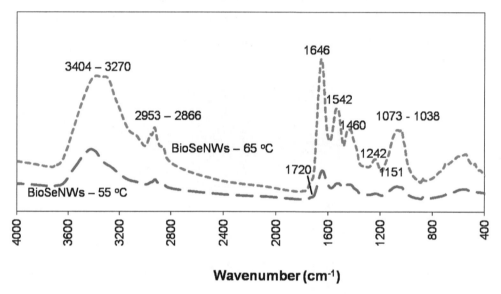

Wavenumber (cm^{-1})

Figure 4.4. FT-IR spectra of BioSeNWs produced at 55 and 65 °C.

The acid-base titration of the BioSeNWs produced at 55 and 65 °C were carried out to determine the pKa values of various functional groups. The slower slope of both the BioSeNWs as compared to Milli-Q water suggests the presence of surface functional groups on the BioSeNWs (data not shown). The local minima in the derivative of the acid-base titration versus pH gives pKa values of the various functional groups as described in an earlier study (Jain et al., 2014a). The local minima were observed at pH 7.3, 5.4 and 3.4 for the BioSeNWs produced at 55 °C. For the BioSeNWs produced at 65 °C, the local minima were observed at 7.2, 5.6 and 3.4. The pKa values observed at 7.3-7.2 can be assigned to sulfonic, sulfinic or thiol groups. The pKa values at 5.6-5.4 can be assigned to phosphoryl or carboxylic acid groups. The pKa values at 3.4 can be assigned to carboxylic acid groups.

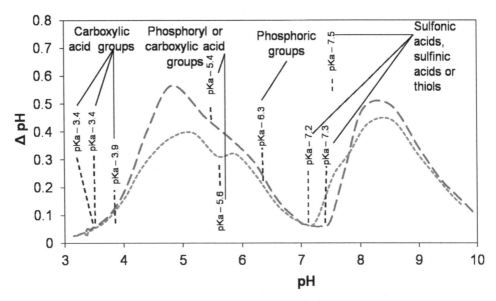

Figure 4.5. Derivative of pH plotted versus pH to determine the pKa values of various functional groups present on the surface of BioSeNWs produced at 55 °C (—) and 65 °C (—).

4.4. Discussion

4.4.1. Morphology of the produced BioSeNWs

This study demonstrates for the first time the biological synthesis of selenium nanowires and further characterized these BioSeNWs. When comparing the aspect ratio (length/diameter) of thermophically produced BioSeNWs and room temperature CheSeNWs, the BioSeNWs showed a similar or better aspect ratio thus suggesting that the produced BioSeNWs are comparable in shape and size to the CheSeNWs. The median diameter of the produced BioSeNWs (20-30 and 40-50 nm) also compares well with the uniform trigonal selenium nanowires produced after refluxing the reduction reaction at 100 °C (Gates et al., 2002).

4.4.2. Surface coating on the BioSeNWs

The produced BioSeNWs are colloidally stable (Figure 3), unlike the CheSeNPs, due to the presence of the capping agent on the BioSeNWs that provides colloidal stability to the nanoparticles through either electrostatic or steric interactinos or both (El Badawy et al., 2010). Indeed, the acid-base titration results and FT-IR confirmed the presence of an organic layer on the surface of BioSeNWs (Figure 4, 5). The FT-IR confirmed the presence of proteins and carbohydrates. The protein and carbohydrates are the main component of EPS and their presence on BioSeNWs suggests that the organic layer capping the BioSeNWs can be the EPS. The ζ-potential vs pH profile of BioSeNWs are very similar to the one of EPS produced by the selenium fed anaerobic granular sludge inoculum (Jain et al., 2014a). Moreover, the iso-electric point of BioSeNWs is very close to the one observed for EPS, EPS capped CheSeNPs and BioSeNPs (Jain et al., 2014a). This suggests that like biogenic elemental selenium nanoparticles, the surface charge of BioSeNWs is governed by the EPS, rather than by the elemental selenium. Thus, the EPS govern the surface charge of formed elemental selenium nanoparticles at all the temperatures.

4.4.3. Formation of BioSeNWs

Microbial reduction of selenite always results in the formation of spherical elemental selenium nanoparticles at ambient temperatures (Jain et al., 2014; Oremland et al., 2004). The produced selenium spherical nanoparticles sometimes transformed to trigonal selenium (Wang et al., 2010) or can stay stable for months (Oremland et al., 2004). In this study, we demonstrated that the trigonal phase of selenium can be easily triggered in-situ by incubating at slightly higher temperatures (55 and 65 °C). This transformation may follow the scheme detailed by Wang et al. (2010), where the reduction of selenite is considered to lead to the formation of amorphous elemental selenium nanspheres, which produce seeds of trigonal selenium which then transform the amorphous selenium to trigonal BioSeNWs. This transformation is attributed to the lower free energy of trigonal elemental selenium as compared to the higher free energy of monoclinic or amorphous elemental selenium (Wang et al., 2010). The glass transition temperature of elemental selenium is 31 °C. Thus, above this temperature, the transformation to trigonal BioSeNWs is favorable. The CheSeNPs formed at room temperature (less than 31 °C) has trigonal crystalline structure (Figure 2). Thus, the transformation of amorphous elemental selenium to trigonal selenium is possible at room temperature. However, this transformation does not readily take place when nanospheres are produced by microbial reduction (Oremland et al., 2004).

The reason for this ambiguity might be due to presence of the extracellular polymeric substance (EPS) present on the surface of biogenic elemental selenium nanospheres (Jain et al., 2014a). Indeed, the presence of EPS has been shown to produce nanospheres as compared to nanowires when the EPS is absent during the chemical reduction of selenite by L-glutathione (Jain et al., 2014a). In the present study, the transformation to nanowires could not be prevented in the presence of EPS. Thus, the role of EPS in controlling the shape and crystallinity is limited to mesophilic temperature conditions, while merely the temperature plays the key role in determining the shape and crystallinity of the biogenic elemental selenium nanoparticles.

4.5. Conclusion

This study demonstrated that colloidal BioSeNWs can be synthesized *in-situ* by microbial reduction of selenite under thermopohilic (55 and 65 °C) conditions. Temperature plays an important role in determining the crystallinity of the BioSeNWs. The produced BioSeNWs were similar in shape, size and crystallinity to CheSeNPs. The colloidal nature of BioSeNWs was due to negative ζ-potential values at near-neutral pH. The negative ζ-potential of BioSeNWs is due to the presence of an organic coating on the surface of the BioSeNWs, most likely originating from the EPS. The surface charge of BioSeNWs is governed by the properties of the EPS rather than those of elemental selenium.

Acknowledgement

The authors are thankful to Dr. Graciella Gil-Gonzalez (KAUST, Saudi Arabia) for insightful discussion, Ferdi Battles (UNESCO-IHE, The Netherlands) for the Nano-sizer and acid-base titration experiments, and Elfi Christalle (Helmholtz-Zentrum Dresden-Rossendorf, Germany) for SEM-EDXS measurements.

4.6. References

Buchs, B., Evangelou, M.W.-H., Winkel, L., Lenz, M., 2013. Colloidal properties of nanoparticular biogenic selenium govern environmental fate and bioremediation effectiveness. Environ. Sci. Technol. 47, 2401–2407.

Dhanjal, S., Cameotra, S.S., 2010. Aerobic biogenesis of selenium nanospheres by *Bacillus cereus* isolated from coalmine soil. Microb. Cell Fact. 9, 52.

El Badawy, A.M., Luxton, T.P., Silva, R.G., Scheckel, K.G., Suidan, M.T., Tolaymat, T.M., 2010. Impact of environmental conditions (pH, ionic strength, and electrolyte type) on the surface charge and aggregation of silver nanoparticles suspensions. Environ. Sci. Technol. 44, 1260–1266.

Faramarzi, M.A., Sadighi, A., 2013. Insights into biogenic and chemical production of inorganic nanomaterials and nanostructures. Adv. Colloid Interface Sci. 189-190, 1–20.

Gates, B., Mayers, B., Cattle, B., Xia, Y., 2002. Synthesis and characterization of uniform nanowires of trigonal Selenium. Adv. Funct. Mater. 12, 219–227.

Jain, R.; Gonzalez-Gil, G.; Singh, V., van Hullebusch, E.D., Farges, F.; Lens, P.N.L., 2014. Biogenic selenium nanoparticles, Production, characterization and challenges. In Kumar, A., Govil, J.N., Eds. Nanobiotechnology. Studium Press LLC, USA, pp. 361-390.

Jain, R., Jordan, N., Schild, D., Hullebusch, E.D. Van, Weiss, S., Franzen, C., Hubner, R., Farges, F., Lens, P.N.L., 2015. Adsorption of zinc by biogenic elemental selenium nanoparticles. Chem. Eng. J. 260, 850–863.

Jain, R., Jordan, N., Weiss, S., Foerstendorf, H., Heim, K., Kacker, R., Hübner, R., Kramer, H., van Hullebusch, E.D., Farges, F., Lens, P.N.L., 2014a. Extracellular polymeric substance govern surface charge of biogenic elemental selenium nanoparticles. Submitted to Environ. Sci. Tech. (Chapter 3 of the present thesis).

Kharissova, O. V, Dias, H.V.R., Kharisov, B.I., Pérez, B.O., Pérez, V.M.J., 2013. The greener synthesis of nanoparticles. Trends Biotechnol. 31, 240–248.

Oremland, R.S., Herbel, M.J., Blum, J.S., Langley, S., Beveridge, T.J., Ajayan, P.M., Sutto, T., Ellis, A. V, Curran, S., 2004. Structural and spectral features of selenium nanospheres produced by Se-respiring bacteria. Appl. Environ. Microbiol. 70, 52–60.

Shah, C.P., Singh, K.K., Kumar, M., Bajaj, P.N., 2010. Vinyl monomers-induced synthesis of polyvinyl alcohol-stabilized selenium nanoparticles. Mater. Res. Bull. 45, 56–62.

Wang, T., Yang, L., Zhang, B., Liu, J., 2010. Extracellular biosynthesis and transformation of selenium nanoparticles and application in H2O2 biosensor. Colloids Surf. B. Biointerfaces 80, 94–102.

Wang, W., Martin, J.C., Fan, X., Han, A., Luo, Z., Sun, L., 2012. Silica nanoparticles and frameworks from rice husk biomass. ACS Appl. Mater. Interfaces 4, 977–981.

Zhang, J.S., Gao, X.Y., Zhang, L.D., Bao, Y.P., 2001. Biological effects of a nano red elemental selenium. Biofactors 15, 27–38.

CHAPTER 5

Understanding the interactions of elemental selenium and amino acids

Abstract:

Understanding the interaction of elemental selenium and amino acids is important for not only designing new self-assembled selenium based nanomaterials but also to determine the fate of colloidal biogenic elemental selenium nanoparticles (BioSeNPs) in the environment. There are few studies that have demonstrated the presence of proteins or peptides on the surface of the BioSeNPs. However, there are no study that has analyzed the interaction of BioSeNPs and peptides neither at a macroscopic nor at a molecular level. This study proposes to reveal the interaction of BioSeNPs and amino acids at microscopic level. The objective of the study is to evaluate the binding preference of amino acids towards elemental selenium. For this, elemental selenium nanoparticles were produced at mesophilic and thermophilic conditions. The proteomics data analysis suggested the presence many proteins including 60 kDa chaperonin, ATP synthase subunits (alpha and beta), chaperone protein dnaK and elongation factor Tu. The statistical analysis on the selected proteins revealed the occurrence of hydrophibic amino acids indicating that the interaction of elemental selenium and amino acids might be of hydrophobic nature. Further research including more critical analysis of proteomic data and density functional theory calculations is necessary to achieve the objectives.

Keywords: amino acids, elemental, selenium, 60 kDa chaperonin, biogenic

Graphical abstract:

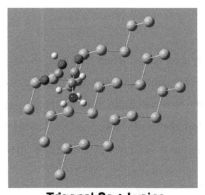

Trigonal Se + Lysine

5.1. Introduction

The interaction of polypeptide with metal(loid) surfaces is studied to design self assembling of nanodevices, to observe cell adhesion to biomaterials or selectivity of a biosensor (Puddu & Perry, 2012). For example, specificity of arginine-rich anti-gold antibodies towards gold is exploited for the self assembly of colloidal gold where antibodies are a selective linker between colloidal gold nanoparticles and bulk gold surface (P. Jain et al., 2014). Understanding the interaction of polypeptides or proteins with elemental selenium is important not only for designing new selenium based nanomaterials but also to determine the formation and fate of colloidal biogenic elemental selenium nanoparticles (BioSeNPs) in bioreactors and environment.

BioSeNPs has been known to be associated with proteins (Debieux et al., 2011; Dobias, Suvorova, & Bernier-latmani, 2011; Lenz, Kolvenbach, Gygax, Moes, & Corvini, 2011). In our previous study, we demonstrated the capping ability of extracellular polymeric substances (EPS), mainly composed of proteins and polysaccharides (Flemming & Wingender, 2010), onto the BioSeNPs (R. Jain et al., 2015). To fully explore the biotechnology for designing of selenium based bottom-up nanodevices or to completely comprehend the factors governing the formation and fate of BioSeNPs in the environment, it is important to understand the interaction of elemental selenium and polypeptides at the molecular level which would imply the study of amino acids or small peptides with elemental selenium (Puddu & Perry, 2012; Ramakrishnan et al., 2013). However, so far there are no study on the interaction of elemental selenium and amino acids.

The objective of this study was to ascertain whether any amino acid bind favorably to the elemental selenium. The evolution of protein also depends on the temperature, thus for this study, elemental selenium nanoparticles were produced at different temperatures. Different temperatures also lead to the formation of different crystal structure of elemental selenium, which may affect the interaction of amino acid with elemental selenium (Ramakrishnan et al., 2013).

5.2 Materials and methods

BioSeNPs were produced and purified under mesophilic (30 °C) and thermophilic conditions (65 °C) as detailed in previous studies (R. Jain et al., 2014, 2015).

The proteomics on the BioSeNPs was carried out as follows. The digested peptide mixtures from proteins in the selenium nanoparticles were obtained following similar protocols as described by Thomas et al. (2013). Digested peptide mixtures were resuspended in 5% (v/v) acetonitrile (ACN) and 0.1% (v/v) formic acid and analyzed on an LTQ Orbitrap Velos mass spectrometer (MS; Thermo Scientific, Bremen, Germany) according to the procedure described by Thomas et al. (2013), except that the flow of the mobile phase was 400 nL min^{-1}. The obtained spectra were submitted to a local MASCOT (version 2.4.0; Matrix Science, London, UK) server and set-up to search bacteria in the Swiss-Prot database (release 2012). Proteins were considered positively identified if the molecular weight search (MOWSE) score from MASCOT was over the 95% confidence limit corresponding to a score ≥ 37. Data validation was performed with Scaffold (version 4.3.2; Proteome software, Portland, OR) at a protein and peptide thresholds of 95%, a minimum of one unique peptide and a protein false discovery rate (FDR) of ≤ 1%. Protein probabilities were assigned by the Protein Prophet algorithm (Nesvizhskii, Keller, Kolker, & Aebersold, 2003). Results from two extractions were combined to increase protein coverage. It is important to note that in the subsequent analysis of the proteomic data, those proteins with less than 4 unique peptides were considered insignificant and thus, discarded.

5.3. Results and discussion

The proteomics data revealed the presence of many different functional proteins as shown in Table 5.1. The major proteins present in both mesophilic and thermophilic selenium nanoparticles were 60 kDa chaperonin, ATP synthase subunits (alpha and beta), chaperone protein dnaK and elongation factor Tu. The presence of these protein were also observed in the study by Lenz et al. (2011). All the above proteins, except elongation factor Tu, have shown preference towards both biogenic and chemogenic elemental selenium (Lenz et al., 2011). Elongation factor Tu and 30S ribosomal protein (S1 and S2) have shown preference towards only biogenic elemental selenium. Other proteins such as lysine-sensitive aspartokinase 3 and Trigger factor have shown selectivity towards chemogenic elemental selenium.

To evaluate the possibility of group of amino acid favoring the interaction with elemental selenium, the four protein that interact with biogenic and chemogenic, two proteins that interact only with biogenic but not with chemogenic and the two more proteins that interact with only chemogenic but not biogenic elemental selenium were chosen. The proteins are listed in Table 5.2.

A simple counting of the particular amino acid in the sequences of the selected proteins and then selecting the top 5 amino acids present in all the selected proteins was used to draw the order for the occurrence of a single amino acid (Table 5.2). The top four amino acids were identified as glycine, valine, leucine and alanine when the data was analyzed for all the selected proteins. It is interesting to note that among the selected amino acids, three of them are hydrophobic. This suggests the possibility of hydrophobic interaction of amino acids and elemental selenium as observed in the case interaction of silica with amino acids(Puddu & Perry, 2012). This finding is also true for the proteins that were showing affinity to both biogenic and chemogenic selenium particles (60 kDa chaperonin, ATP synthase subunits (alpha and beta), chaperone protein dnaK and outer membrane porin protein 32). For the proteins that were showing affinity towards biogenic elemental selenium (elongation factor Tu and 30S ribosomal protein (S1 and S2)), there is a presence of hydrophilic positively charged amino acid (aspartic acid) in the top 4 amino acids. A similar observation was made, but with glutamic acid, for those proteins that were showing affinity to chemogenic elemental selenium (lysine-sensitive aspartokinase 3 and Trigger factor).

5.4. Further work

The above data analysis is the first step towards understanding the interaction of amino acids with elemental selenium. The next steps would include more critical analysis of the data which would include the identification of the active sites in the proteins. This will be followed by the statistical analysis of only those amino acids that are present at the active sites providing indications towards affinity of elemental selenium. The interaction of selenium with the active sites of the proteins will then be studied using density functional theory calculation to determine the affinity of selenium towards amino acids or peptides.

Understanding the interactions of elemental Se and amino acids

Table 5.1. Proteins detected after LTQ Orbitrap and their categorization on the basis of their function. The obtained proteins were also compared with their presence in the earlier study but using pure culture (Lenz et al., 2011). The affinity of the obtained proteins were compared toward biogenic and chemogenic elemental selenium from the literature.

Protein name (Microorganism)	Functions	T = 65 hits*	T = 30 hits	Affinity for	
				Biogenic (Lenz et al., 2011)	Chemogenic
30S ribosomal protein S2 (*Pseudomonas syringae*)	Structural constituent of ribosome	4	0	Y - 30S ribosomal protein S3/S4/S9	N
50S ribosomal protein L1 (*Pseudomonas entomophila*)	primary rRNA binding proteins	6	6	Y	N
50S ribosomal protein L6 (*Pseudomonas fluorescens*)		5	0	Y - 50S ribosomal protein L1	N
60 kDa chaperonin (*Pseudomonas aeruginosa, Bordetella petrii*)	Prevents misfolding and promotes the refolding and proper assembly of unfolded polypeptides generated under stress condition	81	110	Y- many different Chaperonin	Y
Aconitate hydratase 2 (*Pseudomonas aeruginosa*)	Catalyzes the isomerization of citrate to isocitrate	4	6	N	N
Arginine deiminase (*Pseudomonas aeruginosa*)	Catalytic activity	8	11	N	N
ATP synthase subunits (alpha and beta) (*Pseudomonas fluorescens*)	Produces ATP from ADP in the presence of a proton gradient across the membrane.	53	105	Y	Y
Catalase HPII (*Pseudomonas putida*)	Decomposes hydrogen peroxide into water and oxygen	4	6	N	N
Chaperone protein dnaK (*Pseudomonas fluorescens*)	Acts as a chaperone	10	41	Y	Y

	Catalytic activity				
Dihydrolipoamide dehydrogenase (*Pseudomonas putida*)	Protein N(6)-(dihydrolipoyl))lysine + NAD+ = protein N(6)-(lipoyl))lysine + NADH	8	8	Y	N
DNA-directed RNA polymerase subunit alpha (*Pseudomonas fluorescens*)	DNA-dependent RNA polymerase catalyzes the transcription of DNA into RNA	7	6	N	N
Elongation factor Tu 1 (*Methylobacillus flagellatus*)		5	7	Y - Elongation factor TU - *Enterococcus casseliflavis* EC30)	N
Elongation factor Tu 2 (*Acidovorax sp.*)	Promotes the GTP-dependent binding of aminoacyl-tRNA to the A-site of ribosomes during protein biosynthesis	8	11	N	N
Elongation factor Tu (*Pseudomonas fluorescens*)		36	79	Y	N
Lysine-sensitive aspartokinase 3 (*Escherichia coli*)	Catalytic activityi ATP + L-aspartate = ADP + 4-phospho-L-aspartate.	5	0	N - Acetate Kinase and acetate propionate Kinase were associated with Biogenic Se but Aspartate kinase was associated with chemogenic	Y

91

Protein	Function			N - other associated dehydrogenase but not malate dehydrogenas	
Malate dehydrogenase (*Bordetella petri*)	Catalyzes the reversible oxidation of malate to oxaloacetate	4	6	N	N
Maltoporin (*Salmonella agona*)	Involved in the transport of maltose and maltodextrins	4	0	N	N
Outer membrane porin protein 32 (Delftia acidovorans)	Forms anion selective channels.	7	9	Y - Metalloid reductase	Y - Metalloid reducatase
Outer membrane protein A (Bordetella avium)	Structural protein that may protect the integrity of the bacterium.	5	4	N	N
Succinyl-CoA ligase [ADP-forming] subunit alpha and beta (Pseudomonas aeruginosa)	Catalytic activity ATP + succinate + CoA = ADP + phosphate + succinyl-CoA	15	4 (only subunit alpha)	N	N
30S ribosomal protein S1 (Pseudomonas aeruginosa)	Structural constituent of ribosome	0	5	Y - 30S ribosomal protein S3/S4/S9	N
Acetyl-coenzyme A synthetase 1 (Pseudomonas putida)	Catalyzes the conversion of acetate into acetyl-CoA (AcCoA)	0	5	N	N
Acetyl-coenzyme A synthetase (Pseudomonas syringae)		0	5	N	N
Antibiotic efflux pump outer membrane protein arpC (Pseudomonas putida)	Confers resistance to numerous structurally unrelated antibiotics such as carbenicillin, chloramphenicol, erythromycin, novobiocin, streptomycin and tetracycline	0	4	N	N

Understanding the interactions of elemental Se and amino acids

Pyruvate dehydrogenase E1 component (Pseudomonas aeruginosa)	Component of the pyruvate dehydrogenase (PDH) complex, that catalyzes the overall conversion of pyruvate to acetyl-CoA and CO2.	0	5	N	N
Trigger factor OS (Pseudomonas putida)	Involved in protein export. Acts as a chaperone by maintaining the newly synthesized protein in an open conformation	0	4	N	Y

Note: *the "hits" refer to number of times a unique peptide from a particular protein sequence was observed. "T" refers to the temperature of synthesis of biogenic elemental selenium nanoparticles.

Table 5.2. Selected proteins on which the counting of occurrence of amino acids was carried out.

Protein family	Biogenic	Chemogenic	Order of occurrence of amino acids in respective proteins				
60 kDa chaperonin (Pseudomonas aeruginosa)	Y	Y	Alanine	Valine	Glycine	Glutamic acid	Leucine
ATP synthase subunit beta (Pseudomonas fluorescens)	Y	Y	Glycine	Valine	Leucine	Alanine	isoLeucine
ATP synthase subunit alpha OS=Pseudomonas putida	Y	Y	Alanine	Glycine	Valine	Leucine	isoLeucine
Chaperone protein dnaK (Pseudomonas fluorescens)	Y	Y	Alanine	Valine	Aspartic acid	Lysine	Glutamic acid
Outer membrane porin protein 32 (Delftia acidovorans)	Y	Y	Glycine	Alanine	Leucine	Serine	Aspartic acid
30S ribosomal protein S1 (Pseudomonas aeruginosa)	Y	N	Valine	Glutamic acid	Glycine	Aspartic acid	Leucine
30S ribosomal protein S2 (Pseudomonas syringae)	Y	N	Glycine	Alanine	Leucine	isoLeucine	Valine

Elongation factor Tu (Pseudomonas fluorescens)	Y	N	Valine	Glycine	Glutamic acid	Aspartic acid	Alanine
Lysine-sensitive aspartokinase 3 (Escherichia coli)	N	Y	Leucine	Alanine	Valine	Glutamic acid	Glycine
Trigger factor (Pseudomonas putida)	N	Y	Tryptophan	Glycine	Valine	Glutamic acid	Histidine

5.5. References

Debieux, C. M., Dridge, E. J., Mueller, C. M., Splatt, P., Paszkiewicz, K., Knight, I., … Butler, C. S. (2011). A bacterial process for selenium nanosphere assembly. *PNAS*, *108*, 13480–13485.

Dobias, J., Suvorova, E. I., & Bernier-latmani, R. (2011). Role of proteins in controlling selenium nanoparticle size. *Nanotechnology*, *22*(19), 195605.

Flemming, H.-C., & Wingender, J. (2010). The biofilm matrix. *Nature Reviews. Microbiology*, *8*(9), 623–33. doi:10.1038/nrmicro2415

Jain, P., Soshee, A., Narayanan, S. S., Sharma, J., Girard, C., Dujardin, E., & Nizak, C. (2014). Selection of Arginine-Rich Anti-Gold Antibodies Engineered for Plasmonic Colloid Self-Assembly. *The Journal of Physical Chemistry C*, *118*(26), 14502–14510.

Jain, R., Jordan, N., Kacker, R., Weiss, S., Hübner, R., Kramer, H., Hullebusch, E.D. van, Farges, F., Lens, P.N.., 2014a. Biogenic production of trigonal selenium nanowires. (manuscript in preparation, Chapter 4).

Jain, R., Jordan, N., Weiss, S., Foerstendorf, H., Heim, K., Kacker, R., … Lens, P. N. L. (2015). Extracellular Polymeric Substances Govern the Surface Charge of Biogenic Elemental Selenium Nanoparticles. *Environmental Science & Technology*, *49*, 1713–1720.

Lenz, M., Kolvenbach, B., Gygax, B., Moes, S., & Corvini, P. F. X. (2011). Shedding light on selenium biomineralization: proteins associated with bionanominerals. *Applied and Environmental Microbiology*, *77*(13), 4676–80.

Nesvizhskii, A. I., Keller, A., Kolker, E., & Aebersold, R. (2003). A Statistical Model for Identifying Proteins by Tandem Mass Spectrometry abilities that proteins are present in a sample on the basis, *75*(17), 4646–4658.

Puddu, V., & Perry, C. C. (2012). Peptide adsorption on silica nanoparticles: evidence of hydrophobic interactions. *ACS Nano*, *6*(7), 6356–63.

Ramakrishnan, S. K., Martin, M., Cloitre, T., Firlej, L., Cuisinier, F. J. G., & Gergely, C. (2013). Insights on the facet specific adsorption of amino acids and peptides toward platinum. *Journal of Chemical Information and Modeling*, *53*(12), 3273–9.

Thomas, L., Marondedze, C., Ederli, L., Pasqualini, S., & Gehring, C. (2013). Proteomic signatures implicate cAMP in light and temperature responses in Arabidopsis thaliana. *Journal of Proteomics*, *83*, 47–59.

CHAPTER 6

Adsorption of zinc by biogenic elemental selenium nanoparticles

This chapter is accepted for publication as

Jain, R., Jordan, N., Schild, D., van Hullebusch, E.D., Weiss, S., Franzen, C., Hubner, R., Farges, F., Lens, P.N.L., 2015. Adsorption of zinc by biogenic elemental selenium nanoparticles. Chem. Eng. J. 260, 850–863.

Abstract:

The adsorption of Zn^{2+} ions onto biogenic elemental selenium nanoparticles (BioSeNPs) was investigated. BioSeNPs were produced by reduction of selenite (SeO_3^{2-}) in the presence of anaerobic granules from a full scale upflow anaerobic sludge blanket (UASB) reactor treating paper mill wastewater. The BioSeNPs have an iso-electric point at pH 3.8 at 5 mM background electrolyte concentration. X-ray photoelectron spectroscopy showed the presence of a layer of extracellular polymeric substances on the surface of BioSeNPs providing colloidal stability. Batch adsorption experiments showed that the uptake of Zn^{2+} ions by BioSeNPs was fast and occurred at a pH as low as 3.9. The maximum adsorption capacity observed was 60 mg of zinc adsorbed per g of BioSeNPs. The Zn^{2+} ions adsorption on the BioSeNPs was largely unaffected by the presence of Na^+ and Mg^{2+}, but was impacted by the presence of Ca^{2+} and Fe^{2+} ions. The colloidal stability of BioSeNPs decreased with the increasing Zn^{2+} ions loading on BioSeNPs (increase in mg of zinc adsorbed per g of BioSeNPs), corresponding to the neutralization of the negative surface charge of the BioSeNPs, suggesting gravity settling as a technique for solid-liquid separation after adsorption. This study proposes a novel technology for removal of divalent cationic heavy metals by their adsorption on the BioSeNPs present in the effluent of an UASB reactor treating selenium oxyanions containing wastewaters.

Keywords: adsorption, selenium nanoparticles, zinc removal, XPS analysis, ζ-potential, colloidal stability

Graphical abstract:

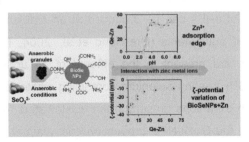

6.1. Introduction

Heavy metals at elevated concentrations are toxic to humans, animals and aquatic ecosystems (Singh et al., 2010). Removal of heavy metals from wastewater is carried out using a variety of techniques, including chemical precipitation, ion exchange, adsorption, membrane filtration or electrochemical separation (Hua et al., 2012). Among these technologies, the adsorption process is advantageous as it is cheap, flexible to operate and maintain, and also generates a high quality effluent, even when metal ions are present in low concentrations in the feed wastewater. Therefore, there is a constant search for adsorbents with higher adsorption capacity, faster kinetics and low cost (Arias and Sen, 2009; Reddad et al., 2002; Wang et al., 2010).

Biologically produced elemental selenium nanoparticles (BioSeNPs) can be a potential new adsorbent for heavy metal cations such as zinc, copper, nickel, lead and cadmium due to the BioSeNPs' amorphous nature (Jain et al., 2014), small diameter (~300 nm (Jain et al., 2014; Oremland et al., 2004) and negative surface charge (ζ-potential −35 mV at neutral pH and 5 mM background electrolyte concentration) [8,9]. Indeed, chemically produced selenium nanoparticles (CheSeNP) adsorb high quantities of copper (800 mg of Cu adsorbed per g of elemental selenium nanoparticles) (Bai et al., 2011). Both CheSeNPs and BioSeNPs adsorb mercury from mercury vapor by forming mercury selenide precipitates (Fellowes et al., 2011; Jiang et al., 2012; Johnson et al., 2008). However, chemical elemental selenium nanoparticles production methods entail high production costs and are not environmental benign due to the use of toxic solvents, high temperature and high pressure (Kharissova et al., 2013; Quintana et al., 2002; Stroyuk et al., 2008). In contrast, BioSeNPs can be produced at an ambient temperature without the use of specialized equipments (Oremland et al., 2004; T. Wang et al., 2010). Moreover, BioSeNPs are present in the effluent of an upflow anaerobic sludge blanket reactor (UASB) treating selenium containing wastewaters due to microbial reduction of selenium oxyanions present in the wastewaters to elemental selenium, thus, further reducing the BioSeNPs' production cost (Jain et al., 2014; Lenz et al., 2009)

Zinc was selected as a model divalent heavy metal ion as it is used extensively in metallurgy, transport, power and construction industries. Zinc is also a major micronutrient in the human body (Hambidge and Krebs, 2007). However, overexposure to zinc can cause stomach cramps, skin irritations, vomiting, anemia, damage to the pancreas, cause arteriosclerosis, impair immune functioning and disturb protein metabolism (Naito et al., 2010). Due to the adverse health impact of overexposure to zinc and the undesirable taste of drinking water at zinc concentration higher than 5 mg L^{-1}, the environmental regulatory agency of the USA and the Food and Agriculture Organization of the United Nations have set limits of 5 mg L^{-1} in the drinking water (USEPA, 2013; FAO, 2014) and the European Commission have set a limit of 5-10 mg L^{-1} zinc in domestic wastewater (EC, 2001). This manuscript, for the first time, attempts to study the interaction of BioSeNPs with Zn^{2+} ions. Based on this fundamental understanding, a zinc removal unit and technique can be developed.

In this study, BioSeNPs were produced by the anaerobic reduction of selenite in presence of anaerobic granules. The reduction of selenite is reported to take place through dissimilatory respiration and the BioSeNPs are mainly formed in the periplasm or extracellularly (Kessi and Hanselmann, 2004; Li et al., 2014). Prior to batch adsorption experiments, BioSeNPs were characterized by X-ray diffraction (XRD), Scanning Electron Microscopy - Energy Dispersive X-ray Spectroscopy analysis (SEM-EDXS), ζ-potential measurements and X-ray photoelectron spectroscopy (XPS). The Qe-Zn (mg of zinc adsorbed per g of BioSeNPs) was determined as a function of adsorption duration, ionic strength, initial metal solution pH and concentrations, and in the presence of competing cations (Na^+, Ca^{2+}, Mg^{2+} and Fe^{2+}) by means of batch experiments. The effect of Zn^{2+} ions adsorption on the BioSeNPs' colloidal stability and residual BioSeNPs concentration in the filtrate was determined with the increase in Qe-Zn. The adsorption of Zn^{2+} ions on the BioSeNPs was characterized using electrophoretic measurements and XPS analysis. For future practical application, removal of Zn^{2+} ions from synthetic wastewater at low pH by the BioSeNPs present in the simulated effluent of an UASB reactor was investigated.

6.2. Materials and methods

6.2.1. BioSeNPs production and purification

BioSeNPs were produced by incubating anaerobic granular sludge (13 g L^{-1} wet weight) in an oxygen-free growth medium at 30 °C and pH 7.3 for 14 days. The growth medium and incubation conditions were applied as these were successfully used for the reduction of selenate using the same inoculum (Lenz et al., 2008). The growth medium contained (in mg L^{-1}): NH$_4$Cl (300.1), CaCl$_2$·2H$_2$O (14.7), KH$_2$PO$_4$ (245.0), Na$_2$HPO$_4$ (283.9) and KCl (245.9). Acid & alkaline trace elements and vitamins were not added to growth medium to avoid their interaction with the formed BioSeNPs. 2.24 g L^{-1} of sodium lactate and 0.86 g L^{-1} of sodium selenite were used, respectively, as carbon and selenium source. This medium was flushed with nitrogen to maintain anaerobic conditions. Anaerobic granular sludge from a full scale UASB reactor used for treating paper mill wastewater in Eerbeek (The Netherlands), described in detail (Roest et al., 2005), was used as inoculum.

The production of elemental selenium was confirmed by the appearance of red colorization of the medium. The supernatant was collected by simple decanting and concentrated by centrifuging (Hermle Z36 HK) at 37,000 g and 4 °C. The pellet was re-suspended in Milli-Q (18MΩ*cm) water and purified by the protocol from Dobias, et al. (2011) with minor modification. Briefly, the concentrated BioSeNPs were sonicated in ice cold water for 1 hour at 100 watt and 42 kHz. NaOH (6 N) was added to raise the pH to 12.5 and the concentrated BioSeNPs were again sonicated at 42 kHz in ice cold water for 2 hours to lyse any remaining biomass present in the supernatant. The pH was lowered back to 7.3 by addition of 1 N HCl. An equal volume of n-hexane was added and the resultant mixture was kept overnight in a separatory funnel. The BioSeNPs were collected from the aqueous phase and washed three times with Milli-Q (18MΩ*cm) water. After washing, the BioSeNPs were re-suspended in Milli-Q water and the pH was adjusted to 7.3 by the addition of a few drops of 1 N NaOH before adding them for adsorption experiments. The BioSeNPs preparation were carried out in duplicate to ensure that the characteristics of the nanomaterial are reproducible.

6.2.2. Batch adsorption experiments

Batch isotherm studies were carried out at different initial zinc concentrations (5.8 - 215.0 mg L^{-1}, 960 minutes of shaking, pH 6.5, added as $ZnCl_2$). Time-dependency studies were carried out at different contact times (1 - 960 minutes, pH 6.5, 70 mg L^{-1} of initial zinc concentration). No background electrolyte was added for isotherm and time-dependency studies. 3 mL of 0.917 g L^{-1} BioSeNPs at pH 7.3 were added as adsorbent. The experiments were carried out at 30 °C, under atmospheric conditions for 16 hours (this duration was found sufficient to reach equilibrium as observed from kinetic experiments). Adsorption of Zn^{2+} ions was also carried out at various initial zinc solution pH (2.0 - 7.2) values with an initial zinc concentration of 70 mg L^{-1}. The theoretical pH values varied from 2.7 to 7.2, and were determined by calculating the final concentration of H^+ ions in the samples while discounting any adsorption reactions. For example, when 3 mL of BioSeNPs at pH 7.3 (H^+ concentration is $10^{-7.3}$ M) was added to 7 mL of zinc metal ion solution of pH 3 (H^+ concentration will be 10^{-3} M), the final H^+ ion concentration and the theoretical pH in the sample will be approximately $7{\times}10^{-4}$ M and 3.2, respectively. The adsorption experiments were also carried out at various ionic strengths (0.7 - 70.0 mM NaCl) and in the presence of competing cations (Ca^{2+}, Mg^{2+} and Fe^{2+}, added as $CaCl_2 2H_2O$, $MgCl_2 6H_2O$ and $FeCl_2 4H_2O$, respectively). The zinc and competing ions were added simultaneously at the beginning of the adsorption experiments. Simulation by Visual MINTEQ software confirmed that more than 98% of the total initial zinc added was in the Zn^{2+} speciation in all the batch adsorption experiments.

As the volumes used in the batch experiments were low, the use of gravity settling for solid-liquid separation was difficult. The samples were, therefore, filtered with a 0.45 µm syringe filter (cellulose acetate, Sigma Aldrich) to be analyzed for the residual zinc concentration. Control experiments were carried out to discard the possibility of adsorption of Zn^{2+} ions to the filter material or by precipitation (more details in Appendix 2). All the experiments were carried out in duplicate. If the difference in two measurements exceeded 10%, experiments were repeated. The average values and errors of duplicate measurements are presented in the figures.

6.2.3. Analysis of BioSeNPs loaded with Zn^{2+} ions

5 mL of BioSeNPs (0.22 g L^{-1}, pH 7.3) was used for adsorbing different initial zinc concentrations (0.9 - 90.9 mg L^{-1}, pH 5.5). The final volume of the samples was 5.5 mL and the final pH of the samples varied between 5.8 and 6.5. The ζ-potential of the BioSeNPs loaded with different concentrations of Zn^{2+} ions was measured (more details in Appendix 2). The samples were then filtered with a 0.45 µm syringe filter (cellulose acetate, Sigma Aldrich) and analyzed for the selenium concentration in the filtrate by ICP-MS. For XPS analysis of BioSeNPs loaded with Zn^{2+} ions, 90.1 mg L^{-1} of zinc were added to BioSeNPs (0.917 g L^{-1}). The final pH of the BioSeNPs loaded with zinc after adsorption was 6.2. Prior to XPS analysis, the samples were centrifuged at 37,000 g for 15 minutes followed by re-suspension in Milli-Q water (see Appendix 2 for details).

6.2.4. Adsorption experiments with simulated wastewaters

Synthetic wastewater containing zinc was generated by adding chloride salts of Zn^{2+} (30 mg L^{-1}), Mg^{2+} (64.6 mg L^{-1}), Ca^{2+} (24 mg L^{-1}) and NH_4^+ (60 mg L^{-1}) as described in Zhao et al. (Zhao et al., 1999), but the pH was adjusted to 2.9 to prove applicability of the proposed technology for more challenging wastewater. The effluent of an UASB reactor containing BioSeNPs was simulated by using effluent of batch incubations without any post treatment (Lenz et al., 2008). This effluent contained Cl^- (766 mg L^{-1}), NO_3^- (29 mg L^{-1}), PO_4^{3-} (50 mg L^{-1}), SO_4^{2-} (159 mg L^{-1}), BioSeNPs (34.2 mg L^{-1}) and 860 mg L^{-1} of Carbon Oxygen Demand. The synthetic zinc containing wastewater was mixed with simulated effluent containing BioSeNPs at a ratio of 1:1 and 1:1.5 for 60 minutes followed by 60 minutes of gravity settling. No filtration was used for solid-liquid separation. After the setting, residual zinc and selenium concentrations were measured in the supernatant. The control experiments were carried out by the same effluent but after removal of BioSeNPs by centrifuging at 37,000 g and collecting the supernatant.

6.2.5. Analytics

Residual zinc, calcium, magnesium and iron concentrations were measured by Atomic Absorption Spectroscopy (see Appendix 2 for details). The selenium content of the BioSeNPs was determined by ICP-MS after being dissolved in concentrated HNO_3 (see Appendix 2 for details). The produced BioSeNPs were characterized by SEM-EDX spectra, XRD, ζ-potential measurements and XPS (more details in Appendix 2). All chemicals were of analytical grade and purchased from Sigma Aldrich (The Netherlands).

6.3. Results

6.3.1. Characterization of the BioSeNPs

The BioSeNPs' particles were spherical in shape (Figure 6.1a) and mainly composed of selenium (Figure 6.1b). In addition, carbon, oxygen as well as small amounts of nitrogen and sulfur were detected by EDX spectra of the SEM (Figure 6.1b). The presence of carbon, oxygen, nitrogen and sulfur can be attributed to the presence of extracellular polymeric substances (EPS) attached to the BioSeNPs, which was further confirmed by the XPS data: the C 1s, N 1s and O 1s peaks were found in the XPS analysis of BioSeNPs (see Figure S1 in Appendix 2). These EPS bound to the surface of the BioSeNPs particles provide colloidal stability to the BioSeNPs at different pH values (Buchs et al., 2013). Note that the large Si signal in Figure 6.1b was due to the use of a silicon wafer during the SEM-EDXS measurements.

The BioSeNPs diameter varied between 80 and 260 nm with a median of 160 - 180 nm (see Figure S2 in Appendix 2). When filtering the BioSeNPs with 1.0 μm and 0.45 μm filters, the filtrate fraction of BioSeNPs obtained was 13.9% and 5.2% of the original concentration, respectively, suggesting retention of the BioSeNPs to the filter. The presence of EPS (Figure 6.1b, S1 in Appendix 2), that can interact with filter material and also increase the hydro-dynamic diameter, can be the cause of this retention (Kayaalp et al., 2014; Nuengjamnong et al., 2005). XRD patterns of BioSeNPs after purification (see Figure S3 in Appendix 2) only show hints for diffuse scattering, suggesting an amorphous nature of the BioSeNPs (Figure S3 in Appendix 2), as observed in previous studies (Tejo Prakash et al., 2009; Wang et al., 2010).

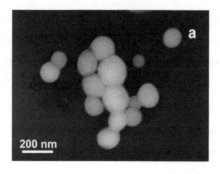

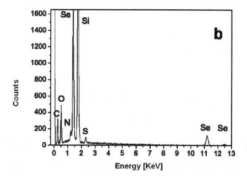

Figure 6.1. a) Secondary electron SEM image of the BioSeNPs deposited onto a piece of Si wafer and b) EDXS analysis of BioSeNPs.

6.3.2. Adsorption of Zn^{2+} ions by BioSeNPs

6.3.2.1. Time-dependency study

The equilibrium pH of the samples was 6.5 at all contact times tested. More than 70% of the Zn^{2+} ions were adsorbed in the first minute of reaction (Figure 6.2a). The uptake of Zn^{2+} ions was completed within 4 hours and remained unchanged for longer contact times. All the further experiments were thus carried out for 16 hours to ensure adsorption equilibrium was achieved.

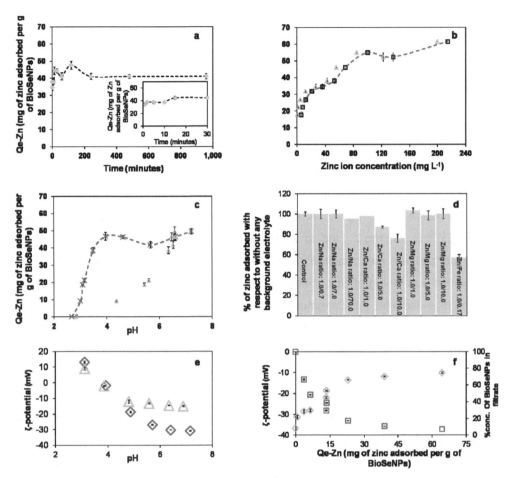

Figure 6.2. Batch adsorption experiment of Zn[2+] ions by BioSeNPs. a) Adsorption at pH 6.5, initial zinc concentration of 70 mg L[−1] and no background electrolyte with time. Zoomed adsorption kinetics for the first 30 minutes presented in the inset. Each data point represents an independent sample; b) Adsorption isotherm at pH 6.5 with different initial (□ with dotted line) and equilibrium (△) zinc concentration; c) Adsorption at 70 mg L[−1] zinc with theoretical pH (× with dotted line) and equilibrium pH (△); d) Adsorption at pH 6.5 and 1.0 mM initial zinc concentration with different competing cations, (e) ζ-potential measurements of BioSeNPs (◇) and BioSeNPs exposed to 1 mM zinc (△) at a background electrolyte concentration of 5 mM NaCl, (f) ζ-potential measurements of BioSeNPs (◊) and % concentration of BioSeNPs in the filtrate (□) after exposure to increasing zinc concentrations. The % of BioSeNPs'

concentration in the filtrate was calculated compared to filtration of BioSeNPs without Zn^{2+} ions adsorption.

6.3.2.2. Adsorption isotherms

The equilibrium pH remained at 6.5 for all the initial zinc concentrations tested. The adsorption of Zn^{2+} ions increased with increase in initial zinc concentration (Figure 6.2b). The adsorption increased sharply when the initial zinc concentration was increased from 5.8 mg L^{-1} to 21 mg L^{-1}. The plateau was reached at an initial zinc concentration of 36 mg L^{-1}. A further increase in initial zinc concentration from 36 mg L^{-1} to 215 mg L^{-1} led to an increase in the adsorption of Zn^{2+} ions and a second plateau was reached. The maximum adsorption capacity of BioSeNPs achieved was 60 mg of zinc adsorbed per g of BioSeNPs. The adsorption isotherms were modeled with the Langmuir and Freundlich models to obtain the theoretical adsorption capacity (Figure S4 in Appendix 2) as explained in Shin et al. (2011). The R^2 values obtained for Langmuir and Freundlich were 0.980 and 0.977, respectively. The Qe-Zn predicted by the Langmuir and Freundlich are 62.1 and 45.5 mg of zinc per g of BioSeNPs. The maximum Qe-Zn predicted by the Langmuir model was close to that observed in the experiments, while the one predicted by the Freundlich model was 25% lower than that observed in the experiments.

6.3.2.3. Effect of variation of pH

In the pH study, the amount of Zn^{2+} ions adsorbed increased with increasing theoretical and equilibrium pH (Figure 6.2c). The Qe-Zn was 21.1 mg g^{-1} (35% of the maximum adsorption) at a theoretical and equilibrium pH value of 3.2 and 5.6, respectively. A steep increase in Qe-Zn to 38.6 mg g^{-1} (64% of the maximum adsorption value) was observed when the theoretical and equilibrium pH value was increased from 3.2 to 3.5 and 5.6 to 6.4, respectively. The maximum adsorption at 70 mg L^{-1} initial zinc concentration was 45 mg g^{-1} (75% of the maximum adsorption), achieved at theoretical pH values above 3.9 and almost equal equilibrium pH value of 6.5 to 6.6.

6.3.2.4 Effect of competing monovalent and divalent cations

NaCl was used as background electrolyte in the Zn/Na ratios: 1.0/0.7, 1.0/7 and 1.0/70 mM/mM to observe their effect on the adsorption of the Zn^{2+} ions onto BioSeNPs (Figure 6.2d). The initial and equilibrium pH values were constant at 6.5. For the 1.0/0.7 and 1.0/7 mM/mM ratio, the adsorption was 100% as compared to the control experiment (without any electrolyte). With the further decrease in the Zn/Na ratio to 1.0/70 mM/mM, the adsorption was 95% of the zinc adsorption in the absence of any background electrolyte.

Ca and Mg were used as competing cations in the Zn/X (where X = Ca or Mg) ratios: 1.0/1.0, 1.0/5.0 and 1.0/10.0 mM/mM to observe potential competing effects on the adsorption of Zn^{2+} ions on BioSeNPs (Figure 6.2d). The initial and equilibrium pH values were 6.5. The Qe-Zn at 1.0 mM (65.4 mg L^{-1}) initial zinc concentration was 45 mg g^{-1} in the absence of any competitive ions (Figure 6.2b). At similar initial experimental conditions, the Qe-Ca (mg of Ca adsorbed per g of BioSeNPs) was 18.7 and 51.7 mg g^{-1}, at respectively, 1 and 5 mM of calcium in the absence of zinc. Similarly, the Qe-Mg (mg of Mg adsorbed per g of BioSeNPs) was 62.6 and 113.8 mg g^{-1}, at respectively, 1 and 5 mM magnesium in the absence of zinc. It is important to note that though the initial experimental conditions were identical for all experiments, the final pH for Mg^{2+}, Ca^{2+} and Zn^{2+} was 8.3, 8.3 and 6.5, respectively.

With calcium as the competing ion in the ratio (Zn/Ca): 1.0/1.0, 1.0/5.0 and 1.0/10.0 mM/mM, the respective Qe-Zn was 98%, 87% and 76% (42.8, 39.2 and 34.2 mg g^{-1}) of the control experiments (45 mg g^{-1}). The presence of Ca^{2+}, thus, decreases the adsorption of Zn^{2+} ions at a Zn/Ca ratio < 1.0/1.0. With $MgCl_2$ as background electrolyte, the Qe-Zn was almost equal to control experiments (45 mg g^{-1}): 104%, 99% and 100% (46.8, 44.6 and 45 mg g^{-1}) at the Zn/Mg ratio of 1.0/1.0, 1.0/5.0 and 1.0/10.0 mM/mM, respectively. Thus, Mg^{2+} does not impact zinc ion adsorption under the applied experimental conditions.

To see the competitive effect of Fe^{2+}, 0.18 mM (10 mg L^{-1}) of Fe^{2+} was added externally at an initial metals ion pH of 4.0 (Zn concentration was 70 mg L^{-1}). The equilibrium pH of this experiment was 5.0. The Qe-Zn in these conditions was reduced to 60% (27 mg g^{-1}) of the maximum (45 mg g^{-1}), showing iron has a strong competitive adsorption effect.

6.3.2.5. Colloidal stability of BioSeNPs

The ζ-potential of the BioSeNPs particles produced by anaerobic granules was −31.0 mV at pH 7 and 5 mM background electrolyte concentration (Figure 6.2e). Similar negative ζ-potential values of BioSeNPs produced by *Bacillus cereus* and *Bacillus selenatarsenatis* were reported in other studies (Buchs et al., 2013; Dhanjal and Cameotra, 2010). The iso-electric point of BioSeNPs particles was at pH 3.8 as compared to 3.5 of selenium nanoparticles produced by *B. selenatarsenatis* (Buchs et al., 2013). The adsorption of zinc on BioSeNPs has led to less negative ζ-potential values (Figure 6.2e, f). No appreciable change in the iso-electric point of BioSeNPs loaded with Zn^{2+} ions was observed. In Figure 6.2f, with the increase in Qe-Zn from 0 to 23.0, ζ-potential values changed from −36.7 to −13.4 mV. When the Qe-Zn value increases to 64.5, the ζ-potential became less negative (−10.2 mV). Figure 6.2f also shows a decrease in concentration of BioSeNPs in the filtrate after adsorption of zinc with the increase in Qe-Zn. The concentration of BioSeNPs decreased by more than 92% (240 µg L^{-1} in the filtrate after adsorption and filtration as compared to 3200 µg L^{-1} in the filtrate after filtration only) with a Qe-Zn of 62 mg of zinc adsorbed per g of BioSeNPs. It is important to note that after adsorption and filtration, 99.9% of the added BioSeNPs were retained in the filter.

6.3.3. XPS analysis

During the XPS analysis of BioSeNPs, the Se 3d$_{5/2}$ binding energy for the BioSeNPs was observed at 55.3 eV (Figure 6.3a). This is in good agreement with binding energy values previously reported in the literature for elemental selenium (see Tables S1 and S2 in Appendix 2). The assignment of the doublet of Se 3d$_{3/2}$ and Se 3d$_{5/2}$ is often overlooked in the literature and authors only refer to Se 3d, probably due to the use of non-monochromatic X-ray excitation. In agreement with observations of Guo and Lu, (1998), the binding energies of the different elemental selenium phases, e.g. amorphous, trigonal or monoclinic are comparable. No additional peaks at binding energies corresponding to other selenium oxidation states were detected (Figure 6.3a). XPS confirmed, therefore, the formation of BioSeNPs particles via selenium(IV) reduction.

At 53.7 eV, a peak corresponding to the Fe 3p elemental line was observed. The presence of Fe is due to the use of anaerobic granular sludge for BioSeNPs production. The total Fe concentration measured after dissolving BioSeNPs in HNO_3 was $5.4 \pm 2.5\%$ (n=4) w/w of the BioSeNPs. As the signals of Fe in XPS were weak, assignment of the oxidation state of Fe was not possible. Since XPS is a surface probing technique, this suggests that most Fe was not present on the surface of the BioSeNPs but entrapped inside the BioSeNPs.

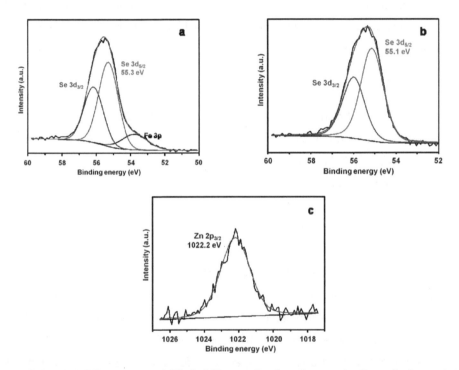

Figure 6.3. (a) XPS spectra of BioSeNPs confirming the production of elemental selenium Se 3d lines of BioSeNPs and XPS spectra of BioSeNPs loaded with zinc, (b) Se 3d lines and (c) Zn $2p_{3/2}$ lines.

The signal of C 1s can be fitted into three components with binding energies located at 284.8, 286.3 and 288.1 eV, corresponding to hydrocarbon chains (C_xH_y), alpha-carbon (α-C) + C-N, and carboxylic acid (COOH groups), respectively (Bansal et al.,

2006, 2005). The N 1s peak is centered at 400.1 eV and lays in the range corresponding to nitrogen containing groups (such as amine or amide groups) (Bansal et al., 2006; Graf et al., 2009; Wang et al., 2010) . The O 1s signal can be fitted into two components at 531.7 and 532.9 eV, corresponding to hydroxyl (—OH) and carboxylate (—COOH) groups, respectively (Senapati et al., 2005) (see Figure S1 in the Appendix 2).

During the XPS analysis of BioSeNPs loaded with zinc, the Se $3d_{5/2}$ binding energy of the Zn^{2+} ions loaded on the BioSeNPs was centered at 55.1 eV (Figure 6.3b), while the Se $3p_{3/2}$ line was found at 161.4 eV. These binding energy values are in agreement with the energies found for the no zinc exposed BioSeNPs, corresponding also to elemental selenium (see Tables S1 and S2 in Appendix 2). No significant differences in the C 1s, N 1s and O 1s lines were observed after interaction with Zn^{2+} ions.

The binding energy of the Zn $2p_{3/2}$ signal is located at 1022.2 eV (Figure 6.3c). The difficulty of attributing this binding energy is due to the fact that Zn compounds such as ZnO, ZnSe, $ZnCO_3$ or $Zn(OH)_2$ (NIST database) show similar Zn $2p_{3/2}$ binding energies (refer to Table S3 in Appendix 2).

6.3.4. Zinc removal from synthetic wastewater

Figure 6.4 demonstrates the zinc removal from simulated zinc containing wastewater. The final pH, after the mixing and settling of the synthetic metal wastewater fed with the simulated UASB effluent containing BioSeNPs and without BioSeNPs, was between 7.6 - 7.8. 97.2±0.2% and 97.2±0.1% of the total zinc was removed at 1:1 and 1:1.5 ratios, respectively, fed with BioSeNPs containing UASB effluent. In the control experiments, 80.7±0.7%, and 79.4±4% of total zinc was removed 1:1 and 1:1.5 ratios, respectively. The enhanced removal of zinc in synthetic wastewaters in comparison to control experiments is due to the presence of BioSeNPs. The removal of zinc in control experiments was due to precipitation of zinc in form of ZnO, $Zn(OH)_2$, $Zn_3(PO4)_2$ and complexes with organic carbon as predicted by Visual MINTEQ. The final zinc concentrations, when BioSeNPs were added, for ratio 1:1 and 1:1.5 were 0.39 and 0.32 mg L^{-1}, respectively, which is well below the regulatory discharge limits (USEPS, 2013; FAO,2014; EC, 2001) and 10 times less than zinc concentration in the control experiments. More than 97 and 80% of added BioSeNPs could be retained in the settled sludge (comprising BioSeNPs and Zn) at the ratio of 1:1 and 1:5, respectively.

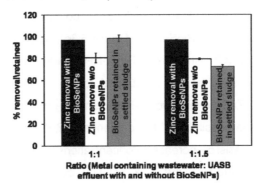

Figure 6.4. Zinc removal efficiency (■) and BioSeNPs retention (■) in the effluent and settled sludge, respectively, at different ratios of synthetic metal wastewater and simulated UASB effluent containing BioSeNPs. Zinc removal efficiency in the control experiments (□) at different ratio when synthetic metal wastewater is mixed with UASB effluent without BioSeNPs.

6.4. Discussion

6.4.1. Mechanisms of Zn^{2+} adsorption onto BioSeNPs at near-neutral pH

This study demonstrated, for the first time, that adsorption of Zn^{2+} ions on BioSeNPs is carried out by different mechanisms depending on the initial zinc concentrations. The ζ-potential of BioSeNPs loaded with Zn^{2+} ions vs Qe-Zn (Figure 6.2f) has the same double-plateau as the adsorption isotherm of Zn^{2+} ions (Figure 6.2b), suggesting two kinds of sorption mechanisms prevail at near neutral pH values. The double-plateau isotherm for Zn^{2+} adsorption on BioSeNPs observed at initial and equilibrium pH of 6.5 (Figure 6.2b) might be due to the presence of high and medium affinity sites on the surface of the adsorbent, as suggested in the adsorption of zinc, cadmium, copper and nickel by amorphous hydrous manganese dioxide (Kanungo et al., 2004) or zinc adsorption by hydroxy intercalated Al and Zr-pillared bentonite (Matthes et al., 1999). Alternatively, the double-plateau adsorption isotherm can be explained by the BET type IV isotherm (Do, 1998). This type of isotherm proposes the formation of adsorbate monolayers on the site, followed by precipitation of adsorbate in the pores of the adsorbent.

XPS analysis of BioSeNPs loaded with Zn^{2+} confirms the formation of zinc precipitates on the surface of BioSeNPs (Figure 6.3c), as predicted in BET type IV isotherm. However, the XPS analysis of BioSeNPs loaded with Zn^{2+} does not allow to assign unambiguously the zinc compound found on the surface, since ZnO, $Zn(OH)_2$, $ZnCO_3$ and ZnSe exhibit very similar Zn $2p_{3/2}$ binding energies (see Table S3 in Appendix 2). The possibility of the presence of ZnO, $Zn(OH)_2$ and $ZnCO_3$ can be explained on the basis of increased concentrations of Zn^{2+} ions in the electrical double layer as compared to the bulk solution due to electrostatic attractions between the high negative ζ-potential of BioSeNPs and the positive charge of the Zn^{2+} ions leading to precipitation of $Zn(OH)_2$, ZnO or $ZnCO_3$ on the surface of the BioSeNPs.

ZnSe could also be present on the surface of BioSeNPs following the disproportionation of elemental selenium to selenium(IV) and selenium($-$II) leading to ZnSe formation (Nuttall, 1987; Su et al., 2000; Zhang et al., 2000). Indeed, based on solubility products, the formation of HgSe and Ag_2Se through disproportionation of elemental selenium into selenide and selenite was reported to be highly favorable, in comparison to ZnSe whose formation was considered to be less favorable (Nuttall, 1987). Such a disproportionation reaction of elemental selenium was experimentally observed during synthesis of CuSe and Ag_2Se in alkaline and hydrothermal conditions (Su et al., 2000), however not at the experimental conditions applied in this study (pH = 6.5, T = 30 °C). Preliminary analysis of Extended X-ray Absorption Fine Structure (EXAFS) data of BioSeNPs loaded with Zn^{2+} ions at Zn K-edge suggests that the first neighbor of Zn is O (See appendix 3). The Zn precipitate can be either ZnO, $Zn(OH)_2$, $ZnCO_3$ or even Zn-organic complexes, rather than ZnSe. To better evaluate the chemical environment of zinc at the BioSeNPs surface, further analysis of the EXAFS data is required, which is beyond the scope of the present study.

6.4.2. Mechanisms of Zn^{2+} ions adsorption onto BioSeNPs at different pH

The adsorption of Zn^{2+} ions on BioSeNPs follows different mechanisms at different solution pH. At initial pH values from 2.9 to 3.8 (the theoretical pH values were calculated to be 3.0 to 4.0 and the equilibrium pH values varied from 4.4 to 6.6), the Zn^{2+} ions adsorption on BioSeNPs followed a ligand-like (type II) adsorption mechanism (Kanungo et al., 2004). In the ligand-like adsorption, ligands can bind to solid surfaces by replacing OH^- and decreasing repulsion between solids and cations, which in turn assist in binding of the cations to the same site as the ligand or at some other sites. Ligands also increase the number of sites taking part in adsorption by maintaining electro-neutrality on the surface of the adsorbent. The excess of OH^- or lack of H^+ ions was observed in the samples while carrying out the mass balance for H^+ ions for the adsorption of Zn^{2+} ions by BioSeNPs at initial pH values of 2.9 to 3.8. The Zn^{2+} ions adsorption was highly correlated to the H^+ sorbed during this pH range ($R^2 = 0.99$, see Figure S5 in Appendix 2). This high correlation suggests that at low pH, a release of OH^- ions or adsorption of H^+ ions takes place during the interaction of Zn^{2+} ions with the BioSeNPs. A similar increase in equilibrium pH was reported during adsorption of Cu^{2+} at initial pH of 3.0 by polyglycidyl methacrylate and polyethyleneimine (Navarro et al., 2001). The equilibrium pH varied between 4.8 and 5.9, increasing with increase of the background chloride ion concentrations, suggesting ligand-like (type II) assisted adsorption (Kanungo et al., 2004).

To quantify the amount of the chloride ions adsorbed, experiments were carried out at an initial zinc concentration of 60.0 mg L^{-1}, chloride ion concentration of 82.8 mg L^{-1}, pH of 3.7 and with 2.2 g L^{-1} of BioSeNPs. The Qe-Cl was 1.8 mg of chloride adsorbed per g of BioSeNPs (see details for Cl^- measurement in Appendix 2). The adsorption of Cl^- points to the possibility of the presence of anion assisted Zn^{2+} ions adsorption by BioSeNPs (Kanungo et al., 2004). Since the Qe-Cl is much lower than that of Qe-Zn (25.5 mg g^{-1}) at these experimental conditions, anion assisted Zn^{2+} ions adsorption is most likely not the dominant mechanism or is only valid for a small pH range.

At the theoretical pH value of 7.2, the replacement of the H^+ ion by Zn^{2+} ions on the surface of BioSeNPs was suggested by the drop in the equilibrium pH to 6.5 from theoretical pH of 7.2. This release of H^+ ion was also observed when the theoretical pH value was increased from 4.0 to 5.7.

6.4.3. Effect of competing ions on Zn^{2+} ions adsorption

The effect of Zn^{2+} ions adsorption on BioSeNPs in the presence of common competing ions as Na, Ca, Mg and Fe is important to assess the applicability of BioSeNPs for real wastewaters. To this point, it was observed that the increase in NaCl concentration from 0.001 M to 0.1 M, there was no significant effect on the zinc ion adsorption by BioSeNPs (Figure 6.2d). This suggests that the Zn^{2+} ions are adsorbed on the surface of BioSeNPs via inner sphere complexation (Wang et al., 2013).

The relative increase in adsorption of the cations either follows a decrease in ionic radius or an increase in electronegativity of the metal ion or an increase in ratio of the ionisation potential and ionic radius (McKay and Porter, 1997). Qe-Mg > Qe-Zn > Qe-Ca, which is explained by the trend in ionic radius of the ions: $Mg^{2+} < Zn^{2+} < Ca^{2+}$ (see Table S4 in Appendix 2). However, the relative preference of cations for adsorption by BioSeNPs follows the trend in the ratio of ionisation potential and ionic radius, which is the highest for Zn^{2+} (−1.03), lower for Ca^{2+} (−2.89) and the lowest for Mg^{2+} (−3.63), thus implying that Zn^{2+} would outcompete calcium and magnesium at the equimolar ratios.

The ratio of ionization potential and ionic radius for Fe^{2+} (−0.77) exceeds that of Zn^{2+} (−1.03), thus Fe^{2+} would outcompete zinc at equimolar ratio. The effect of entrapped Fe in BioSeNPs on its adsorption capacity could not be measured as it is impossible to remove Fe entirely from the BioSeNPs without altering or destroying them. The entrapped Fe is, however, unlikely to have inhibited the adsorption of Zn^{2+} on BioSeNPs as the majority of Fe was not present on the surface of BioSeNPs (Figure 6.3a).

6.4.4. Colloidal stability of BioSeNPs

The increase of Zn^{2+} ions adsorption with increasing pH can be ascribed to the change in ζ-potential of the BioSeNPs. The ζ-potential of the BioSeNPs is negative at pH values above the iso-electric point (pH 3.8) and becomes more negative with increasing pH. The negative charge on the surface of BioSeNPs attracts the Zn^{2+} ions and thus, the more negative charge, the stronger will be the attraction and hence increases the adsorption (Arias and Sen, 2009). Furthermore, the change in pH leads to deprotonation of functional groups present on the surface of BioSeNPs (see Figure S1 in Appendix 2) which, in turn, provides more binding sites to Zn^{2+} ions and thus increases adsorption.

The interaction of Zn^{2+} ions and BioSeNPs leads to less negative BioSeNPs loaded with zinc, suggesting that the zinc is adsorbed either by electrostatic interactions or by covalent bond formation (Figures 6.2e, f and Table S5 in Appendix 2). The same trend was observed for interaction of BioSeNPs with Ca^{2+} and Mg^{2+} (see Figure S6 in Appendix 2) and was also observed during the interaction of calcium ions with BioSeNPs produced by *Bacillus selenatarsenatis* (Buchs et al., 2013). No appreciable shift in iso-electric point of BioSeNPs loaded with zinc was observed (Figure 6.2e). This can be attributed to the relatively small amount of zinc adsorption at an equilibrium pH value of 4. A similar observation was made during adsorption of U(VI) on MnO_2 (Wang et al., 2013).

The ζ-potential becomes less negative at increasing Qe-Zn (Figure 6.2f, Table S5 in Appendix 2). This suggests that as the load of Zn^{2+} ions on the surface of BioSeNPs increases, the colloidal stability of BioSeNPs decreases. The effect of lower colloidal stability of BioSeNPs loaded with Zn^{2+} ions led to settling and a lower concentration of BioSeNPs in the filtrate: 99.9% of retention of total added BioSeNPs on the filter was achieved (240 µg L^{-1} of selenium concentration after zinc adsorption and filtration; 3200 µg L^{-1} after only filtration, 220,000 µg L^{-1} of added BioSeNPs, Figure 6.2f). Thus, the heavy metals loaded on BioSeNPs can be separated from the liquid phase by simple gravity settling.

6.4.5. Practical implications

This study demonstrated that the adsorption of Zn^{2+} ions can be performed at acidic pH values (pH 3.9). This is an interesting characteristic as the pH of the metal containing wastewaters such as electroplating industry wastewater or acid mine drainage wastewater (Lenz et al., 2008) varies from highly acidic to near neutral (Amer, 1998; Boricha and Murthy, 2009; Kanawade and Gaikwad, 2011; Zhao et al., 1999), where activated carbon is unable to adsorb zinc (Kouakou et al., 2013). Though the adsorption capacity of BioSeNPs is low in comparison to Dowex HCR S/S resin (Qe 172.2 mg g^{-1}) (Zhang et al., 2010), NaA and NaX zeolites (Qe 118.9 and 106.4 mg g^{-1}) (Nibou et al., 2010), the adsorption capacity of BioSeNPs is higher than most of the common adsorbents used for zinc removal such as aluminosilicates, non-modified zeolite, bentonite or activated carbon (see Table 6.1).

Table 6.1. Maximum Qe-Zn of common adsorbents for Zn^{2+} ions at relevant conditions

Adsorbent	Q max (mg g^{-1})	pH	Concentration (mg L^{-1}) i = initial conc. e = equilibrium conc.	Reference
BioSeNPs	60	6.5	200 (i)	This study
Hydrous manganese dioxide	57.2, 85.0	6.0	18.3 (e) 3.3 (e)	(Kanungo et al., 2004; Su et al., 2010)
Graphene oxide	345	5.0	100 (e)	(Sitko et al., 2013)
Aluminosilicates	6.5	6.5	65.4 (i)	(Miyazaki et al., 2003)
Al$_2$O$_3$	22.9	6.5	195.9(i)	(Miyazaki et al., 2003)
PVA/EDTA resin	38.7	6.0	40 (i)	(Zhang et al., 2010)

Dowex HCR S/S resin	172.2	6.0	18 (e)	(Zhang et al., 2010)
Sodium rich bentonite	23.6	6.9	97.3 (e)	(Matthes et al., 1999)
Al-pillared Na-rich bentonite	31.8	6.9	97.3 (e)	(Matthes et al., 1999)
Commercial activated powder carbon	20.5	7.0	400 (e)	(Kouakou et al., 2013)

This study opens perspectives to develop a novel adsorption technology where BioSeNPs present in the effluent of an UASB reactor treating selenium oxyanions containing wastewater (Buchs et al., 2013; Lenz et al., 2008) is used as a metal adsorbent. It is important to note that the BioSeNPs are always present in the fore mentioned effluent due to microbial conversion of dissolved selenium oxyanions to BioSeNPs and these BioSeNPs has to be removed prior to discharging of the effluent (Buchs et al., 2013). Figure 6.4 demonstrated that the regulatory discharge concentration of zinc can be achieved by use of BioSeNPs at the tested conditions. It was also observed that at the appropriate metal to BioSeNPs ratio, 1:1 in this study, more than 97% of BioSeNPs can be retained in the settled sludge or removed from the effluent of an UASB reactor treating selenium oxyanion wastewater by a simple cost-effective gravity settling. The settled BioSeNPs loaded with zinc metal then can be used for recovery of the heavy metal and BioSeNPs.

6.5. Conclusion

In this study, the adsorption of Zn^{2+} ions on BioSeNPs was investigated. Adsorption of Zn^{2+} ions on BioSeNPs follows a two-step process at near-neutral pH values and follows ligand-like (type II) mechanisms at acidic pH. Furthermore, Zn^{2+} ions adsorbs to BioSeNPs mainly through inner-sphere complexation. Major advantages of using BioSeNPs as an adsorbent are the material's fast kinetics and capacity to adsorb more than 75% of the maximum adsorption capacity even at pH values below 3.9. The ζ-potential of BioSeNPs changed from −31 mV to −15 mV after interaction with the Zn^{2+} ions, leading to aggregation of the BioSeNPs and subsequent settling of the colloidal suspension. This allows recovering the metal loaded BioSeNPs by simple gravity settling as observed in experiments with synthetic zinc wastewaters. This study provides understanding of Zn^{2+} ions adsorption onto BioSeNPs, which can be exploited to develop a new heavy metal removal process based on BioSeNPs produced out of effluents of UASB reactors treating selenium oxyanions rich wastewaters.

Acknowledgments

The authors are thankful to Dr. Pakshi Rajan Kanan (IIT Guwahati, India) and Domician Dominic (UNESCO-IHE, Delft) for helping with the experiments, Ferdi Battles, Lyzette Robbemont, Don van Galen (UNESCO-IHE, Delft) for the ICP-MS measurements, Marc Strampraad and Shaak Lispeth (TU-Delft, The Netherlands) for allowing access to high speed centrifuge, Gilles Catillon for XRD analysis (Université Paris-Est, Marne-la-Vallée), and Elfi Christalle (Helmholtz-Zentrum Dresden-Rossendorf) for SEM-EDXS measurements.

6.6. References

Amer, S.I., 1998. Treating metal finishing wastewater. Environmental Technology, Aquachem Inc: Missouri City, TX

Arias, F., Sen, T.K., 2009. Removal of zinc metal ion (Zn^{2+}) from its aqueous solution by kaolin clay mineral: A kinetic and equilibrium study. Colloids Surfaces A Physicochem. Eng. Asp. 348, 100–108.

Bai, Y., Rong, F., Wang, H., Zhou, Y., Xie, X., Teng, J., 2011. Removal of copper from aqueous solution by adsorption on elemental selenium nanoparticles. J. Chem. Eng. Data 56, 2563–2568.

Bansal, V., Poddar, P., Ahmad, A., Sastry, M., 2006. Room-Temperature Biosynthesis of Ferroelectric Barium Titanate Nanoparticles 11958–11963.

Bansal, V., Rautaray, D., Bharde, A., Ahire, K., Sanyal, A., Ahmad, A., Sastry, M., 2005. Fungus-mediated biosynthesis of silica and titania particles. J. Mater. Chem. 15, 2583.

Boricha, A.G., Murthy, Z.V.P., 2009. Preparation, characterization and performance of nanofiltration membranes for the treatment of electroplating industry effluent. Sep. Purif. Technol. 65, 282–289.

Buchs, B., Evangelou, M.W.-H., Winkel, L., Lenz, M., 2013. Colloidal properties of nanoparticular biogenic selenium govern environmental fate and bioremediation effectiveness. Environ. Sci. Technol. 47, 2401–2407.

Dhanjal, S., Cameotra, S.S., 2010. Aerobic biogenesis of selenium nanospheres by Bacillus cereus isolated from coalmine soil. Microb. Cell Fact. 9, 52.

Do, D.D., 1998. Adsorption Analysis: Equilibria and Kinetics, Series on Chemical Engineering. Imperial College Press, UK.

Dobias, J., Suvorova, E.., Bernier-latmani, R., 2011. Role of proteins in controlling selenium nanoparticle size. Nanotechnology 22, 195605.

Environment Directorate-General - European Commission: Luxembourg, 2001. Pollutants in urban wastewater and sewage sludge, ISBN 92-894-1735-8, http://ec.europa.eu/environment/waste/sludge/pdf/sludge_pollutants.pdf

Fellowes, J.W., Pattrick, R.A.D., Green, D.I., Dent, A., Lloyd, J.R., Pearce, C.I., 2011. Use of biogenic and abiotic elemental selenium nanospheres to sequester elemental mercury released from mercury contaminated museum specimens. J. Hazard. Mater. 189, 660–669.

Food and Agriculture Organization of the United Nations, 2014. FAO Corporate Document Repository, http://www.fao.org/docrep/x5624e/x5624e04.htm (access date: 14/09/2014)

Graf, N., Yegen, E., Gross, T., Lippitz, A., Weigel, W., Krakert, S., Terfort, A., Unger, W.E.S., 2009. XPS and NEXAFS studies of aliphatic and aromatic amine species on functionalized surfaces. Surf. Sci. 603, 2849–2860.

Guo, F.Q., Lu, K., 1998. Microstructural evolution of an as-quenched amorphous Se sample. Phys. Rev. B 57, 10414.

Hambidge, K.M., Krebs, N.F., 2007. Zinc deficiency: a special challenge. J. Nutr. 137, 1101–1105.

Hua, M., Zhang, S., Pan, B., Zhang, W., Lv, L., Zhang, Q., 2012. Heavy metal removal from water/wastewater by nanosized metal oxides: a review. J. Hazard. Mater. 211-212, 317–331.

Jain, R.; Gonzalez-Gil, G.; Singh, V., van Hullebusch, E.D., Farges, F.; Lens, P.N.L., 2014. Biogenic selenium nanoparticles, Production, characterization and challenges. In Kumar, A., Govil, J.N., Eds. Nanobiotechnology. Studium Press LLC, USA, pp. 361-390.

Jiang, S., Ho, C.T., Lee, J.-H., Duong, H. Van, Han, S., Hur, H.-G., 2012. Mercury capture into biogenic amorphous selenium nanospheres produced by mercury resistant Shewanella putrefaciens 200. Chemosphere 87, 621–4.

Johnson, N.C., Manchester, S., Sarin, L., Gao, Y., Kulaots, I., Hurt, R.H., 2008. Mercury vapor release from broken compact fluorescent lamps and in situ capture by new nanomaterial sorbents. Environ. Sci. Technol. 42, 5772–5778.

Kanawade, S.M., Gaikwad, R.W., 2011. Removal of zinc ions from industrial effluent by using cork powder as adsorbent. Int. J. Chem. Eng. Appl. 2, 199–201.

Kanungo, S.B., Tripathy, S.S., Mishra, S.K., Sahoo, B., 2004. Adsorption of Co^{2+}, Ni^{2+}, Cu^{2+}, and Zn^{2+} onto amorphous hydrous manganese dioxide from simple (1–1) electrolyte solutions. J. Colloid Interface Sci. 269, 11–21.

Kayaalp, N., Kinaci, C., Dizge, N., Hamidi, N., 2014. Correlation of filtration resistance with microbial polymeric substances extracted from membranes in a submerged membrane bioreactor. CLEAN - Soil, Air, Water 42, 1-9

Kessi, J., Hanselmann, K.W., 2004. Similarities between the abiotic reduction of selenite with glutathione and the dissimilatory reaction mediated by *Rhodospirillum rubrum* and *Escherichia coli*. J. Biol. Chem. 279 50662–50669

Kharissova, O. V, Dias, H.V.R., Kharisov, B.I., Pérez, B.O., Pérez, V.M.J., 2013. The greener synthesis of nanoparticles. Trends Biotechnol. 31, 240–248.

Kouakou, U., Ello, A.S., Yapo, J.A., Trokourey, A., 2013. Adsorption of iron and zinc on commercial activated carbon 5, 168–171.

Lenz, M., Hullebusch, E.D. Van, Hommes, G., Corvini, P.F.X., Lens, P.N.L., 2008. Selenate removal in methanogenic and sulfate-reducing upflow anaerobic sludge bed reactors. Water Res. 42, 2184–2194.

Lenz, M., Van Aelst, A.C., Smit, M., Corvini, P.F.X., Lens, P.N.L., 2009. Biological production of selenium nanoparticles from waste waters. Mater. Res. 73, 721–724.

Li, D.-B., Cheng, Y.-Y., Wu, C., Li, W.-W., Li, N., Yang, Z.-C., Tong, Z.-H., Yu, H.-Q., 2014. Selenite reduction by *Shewanella oneidensis* MR-1 is mediated by fumarate reductase in periplasm. Sci. Rep. 4, 3735.

Life, A., Table, C., 2013. National Recommended Water Quality Criteria 1–11.

Matthes, W., Madsen, F.W., Kahr, G., 1999. Sorption of heavy met;s cations by Al and Zr-hydroxyl-intercalated and pillared bentonite Clays and clays minerals 47, 617–629.

McKay, G., Porter, J.F., 1997. Equilibrium Parameters for the Sorption of Copper, Cadmium and Zinc Ions onto Peat. J. Chem. Technol. Biotechnol. 69, 309–320.

Miyazaki, A., Balint, I., Nakano, Y., 2003. Solid-liquid interfacial reaction of Zn^{2+} ions on the surface of amorphous aluminosilicates with various Al/Si ratios. Geochim. Cosmochim. Acta 67, 3833–3844.

Naito, W., Kamo, M., Tsushima, K., Iwasaki, Y., 2010. Exposure and risk assessment of zinc in Japanese surface waters. Sci. Total Environ. 408, 4271–84.

Navarro, R.R., Tatsumi, K., Sumi, K., Matsumura, M., 2001. Role of anions on heavy metal sorption of a cellulose modified with poly(glycidyl methacrylate) and polyethyleneimine. Water Res. 35, 2724–30.

Nibou, D., Mekatel, H., Amokrane, S., Barkat, M., Trari, M., 2010. Adsorption of Zn2+ ions onto NaA and NaX zeolites: kinetic, equilibrium and thermodynamic studies. J. Hazard. Mater. 173, 637–46.

NIST database, n.d.

Nuengjamnong, C., Hyang, J., Cho, J., Polprasert, C., Ahn, K., 2005. Membrane fouling caused by extracellular polymeric substances during microfiltration processes 179, 117–124.

Nuttall, K.L., 1987. A model for metal selenide formation under biological conditions. Medical Hypethesis 24, 217–221.

Oremland, R.S., Herbel, M.J., Blum, J.S., Langley, S., Beveridge, T.J., Ajayan, P.M., Sutto, T., Ellis, A. V, Curran, S., 2004. Structural and spectral features of selenium nanospheres produced by Se-respiring bacteria. Appl. Environ. Microbiol. 70, 52–60.

Quintana, M., Haro-poniatowski, E., Morales, J., Batina, N., 2002. Synthesis of selenium nanoparticles by pulsed laser ablation. Appl. Surf. Sci. 195, 175-186.

Reddad, Z., Gerente, C., Andres, Y., Le Cloirec, P., 2002. Adsorption of several metal ions onto a low-cost biosorbent: kinetic and equilibrium studies. Environ. Sci. Technol. 36, 2067-2073

Roest, K., Heilig, H.G.H.J., Smidt, H., de Vos, W.M., Stams, A.J.M., Akkermans, A.D.L., 2005. Community analysis of a full-scale anaerobic bioreactor treating paper mill wastewater. Syst. Appl. Microbiol. 28, 175–85.

Senapati, S., Ahmad, A., Khan, M.I., Sastry, M., Kumar, R., 2005. Extracellular biosynthesis of bimetallic Au-Ag alloy nanoparticles. Small 1, 517–20.

Shin, K.-Y., Hong, J.-Y., Jang, J., 2011. Heavy metal ion adsorption behavior in nitrogen-doped magnetic carbon nanoparticles: isotherms and kinetic study. J. Hazard. Mater. 190, 36–44.

Singh, A., Sharma, R.K., Agrawal, M., Marshall, F.M., 2010. Health risk assessment of heavy metals via dietary intake of foodstuffs from the wastewater irrigated site of a dry tropical area of India. Food Chem. Toxicol. 48, 611–619.

Sitko, R., Turek, E., Zawisza, B., Malicka, E., Talik, E., Heimann, J., Gagor, A., Feist, B., Wrzalik, R., 2013. Adsorption of divalent metal ions from aqueous solutions using graphene oxide. Dalton Trans. 42, 5682–5689.

Stroyuk, A.L., Raevskaya, A.E., Kuchmiy, S.Y., Dzhagan, V.M., Zahn, D.R.T., Schulze, S., 2008. Structural and optical characterization of colloidal Se nanoparticles prepared via the acidic decomposition of sodium selenosulfate 320, 169–174.

Su, H., Xie, Y., Li, B., Qian, Y., 2000. A simple , convenient , mild hydrothermal route to nanocrystalline CuSe and Ag_2Se. Mat. Res. Bull. 35, 465–469.

Su, Q., Pan, B., Wan, S., Zhang, W., Lv, L., 2010. Use of hydrous manganese dioxide as a potential sorbent for selective removal of lead, cadmium, and zinc ions from water. J. Colloid Interface Sci. 349, 607–612.

Tejo Prakash, N., Sharma, N., Prakash, R., Raina, K.K., Fellowes, J., Pearce, C.I., Lloyd, J.R., Pattrick, R. a D., 2009. Aerobic microbial manufacture of nanoscale selenium: exploiting nature's bio-nanomineralization potential. Biotechnol. Lett. 31, 1857–1862.

United States Environmental Protection Agency website, 2013. National recommended water quality criteria. http://water.epa.gov/drink/contaminants/secondarystandards.cfm.   (access   date: 14/09/2014).

Wang, J., Zheng, S., Shao, Y., Liu, J., Xu, Z., Zhu, D., 2010. Amino-functionalized $Fe_3O_4@SiO_2$ core-shell magnetic nanomaterial as a novel adsorbent for aqueous heavy metals removal. J. Colloid Interface Sci. 349, 293–299.

Wang, T., Yang, L., Zhang, B., Liu, J., 2010. Extracellular biosynthesis and transformation of selenium nanoparticles and application in H_2O_2 biosensor. Colloids Surf. B. Biointerfaces 80, 94–102.

Wang, Z., Lee, S.-W., Catalano, J.G., Lezama-Pacheco, J.S., Bargar, J.R., Tebo, B.M., Giammar, D.E., 2013. Adsorption of Uranium(VI) to Manganese Oxides:

X-ray Absorption Spectroscopy and Surface Complexation Modeling. Environ. Sci. Technol. 47, 850–858.

Zhang, W., Yang, Z., Liu, J., Hui, Z., Yu, W., 2000. A hydrothermal synthesis of orthorhombic nanocrystalline cobalt diselenide CoSe$_2$ Mat. Res. Bull. 35, 2403–2408.

Zhang, Y., Li, Y., Yang, L., Ma, X., Wang, L., Ye, Z., 2010. Characterization and adsorption mechanism of Zn2+ removal by PVA/EDTA resin in polluted water. J. Hazard. Mater. 178, 1046–1054.

Zhao, M., Duncan, J.., van Hille, R.., 1999. Removal and recovery of zinc from solution and electroplating effluent using *Azolla filiculoides*. Water Res. 33, 1516–1522.

CHAPTER 7

Selective adsorption of Cu in a multimetal mixture onto biogenic elemental selenium nanoparticles

This chapter will be submitted as

Jain, R., Dominic, D., Jordan, N., Rene, E.R., Weiss, S., Hullebusch, E.D. Van, Foerstendorf, H., Heim, K., Hubner, R., Farges, F., Lens, P.N.L., 2014. Selective adsorption of Cu in a multimetal mixture onto biogenic elemental selenium nanoparticles. Chem. Eng. J. *(To be submitted)*

Abstract:

Selective adsorption of heavy metals contained in wastewaters is desirable as the selected metal can then be reprocessed and reused. In this study, biogenic elemental selenium nanoparticles (BioSeNPs) were assessed for their ability to selectively adsorb Cu, Cd and Zn from an equimolar mixture. BioSeNPs showed the following preference order for adsorption: Cu>Zn>Cd at the theoretical pH variation of 3.0 to 5.6. BioSeNPs adsorbed 4.7 times more Cu from an equimolar mixture (0.5 mM) of Cu, Zn and Cd at a theoretical pH of 4.3 and metals to BioSeNPs ratio of 1:1 (v:v). Infrared spectroscopy analysis revealed that the Cu, Cd and Zn were interacting with the hydroxyl and carboxyl functional groups present on the surface of the BioSeNPs. The adsorption of the heavy metals onto the BioSeNPs led to less negative ζ-potentials of BioSeNPs, leading to a lower colloidal stability and induced their settling. Cu was the most effective metal in neutralizing the negative surface charge of the BioSeNPs and thus induced a better settling or better retention in the filters.

Keywords: Adsorption, selenium, nanoparticles, BioSeNPs, selective, heavy metals, ζ-potential

Graphical abstract:

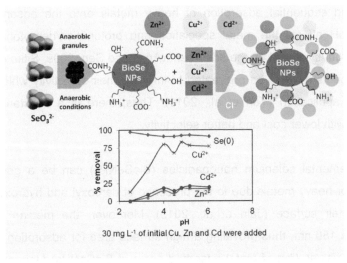

30 mg L^{-1} of initial Cu, Zn and Cd were added

7.1. Introduction

Heavy metals such as Cu, Zn and Cd have many technical and industrial applications. Toxicity induced by exposure to high Cu, Zn and Cd concentrations in humans, animals, plants and the environment is well documented. Thus, regulatory agencies have set discharge limits for these metals, e.g. the United States Environmental Protection Agency (US EPA) limits the maximum contamination level (MCL) to 1.3, 0.5 and 5.0 mg L^{-1} for Cu, Cd and Zn, respectively, in discharged wastewaters (US EPA, 2013). Adsorption is the most recommended technology for the removal of heavy metals, including Cu, Zn and Cd, from low concentration and high volume wastewater, such as acid mine drainage or electroplating industry (Fu and Wang, 2011; Hua et al., 2012; Kanawade and Gaikwad, 2011; Wingenfelder et al., 2005). To separate, reprocess and reuse these metals, their selective adsorption from multimetal mix wastewaters is required.

Selective adsorption of heavy metals onto the adsorbent depends on the metals' properties, such as ionic radius, electronegativity, ratio of ionic radius and ionization potential, as well as metal speciation (McKay and Porter, 1997). The presence of functional groups such as carboxyl and hydroxyl on the surface of adsorbents also leads to selective adsorption of heavy metals (Sitko et al., 2013; Yan et al., 2011). Furthermore, the wastewater pH also plays a significant role in optimizing the selective and sequential adsorption of heavy metals onto the adsorbent as the change in pH affects the metal speciation and protonation/deprotonation of the functional groups present on the adsorbent (Arias and Sen, 2009; Sitko et al., 2013). Most adsorbents have to be pretreated to improve their selectivity which increases the cost of application (Yan et al., 2011). Thus, there is a constant search for adsorbents with lower cost and better selectivity.

Biogenic elemental selenium nanoparticles (BioSeNPs) can be a good selective adsorbent for heavy metals due to the presence of carboxyl and hydroxyl functional group on their surface (Jain et al., 2015). Moreover, the median diameter of BioSeNPs is 180 nm, thus providing a high surface area for adsorption, resulting in high Qe-Me values (mg of metal adsorbed per g of BioSeNPs) (Jain et al., 2015) . Also, BioSeNPs are a low cost adsorbent, as they can be produced by microbial reduction from selenium oxyanions containing wastewaters (Lenz et al., 2009, 2008). Although, there is a single metal adsorption study on the removal of Zn by BioSeNPs (Jain et al., 2015), there are so far no studies carried out on multiple heavy metal adsorption onto BioSeNPs.

The objective of this study was to assess the selective adsorption of Cu, Cd and Zn onto BioSeNPs as well as their effect on the colloidal stability of BioSeNPs. In order to achieve this objective, adsorption of Cu and Cd was performed in a single metal system at different initial pH values. Competitive adsorption of Cu, Cd and Zn at different metals to BioSeNPs ratios was carried out at different pH. The concentration of selenium in the filtrate was measured for all the batch adsorption experiments. The interaction of BioSeNPs with metals was also studied with Scanning Electron Microscopy - Energy Disperse X-ray Spectroscopy (SEM-EDXS), Fourier Transform Infrared Spectroscopy (FT-IR) and zetametry.

7.2. Materials and methods

7.2.1. Production, purification and characterization of BioSeNPs

BioSeNPs were produced, purified and characterized as described in previous study (Jain et al., 2015). Briefly, reduction of selenite was carried out under anaerobic conditions in the presence of anaerobic granular sludge. The produced BioSeNPs were decanted and concentrated by centrifugation at 37,000g. The pellet of BioSeNPs was re-suspended in Milli-Q water (18 MΩ*Cm), followed by sonication, NaOH treatment, hexane separation and finally washing with Milli-Q water.

7.2.2. Batch adsorption tests

Single metal batch adsorption experiments were carried out for Cu (added as $CuCl_2$) and Cd (added as $CdCl_2$). 3 mL of 0.84 g L^{-1} of BioSeNPs (pH 7.3) was added to 7 mL of heavy metal solution. Time-dependent adsorption experiments were carried out for different duration (1 - 960 minutes) at an initial metal ion concentration of 70 mg L^{-1}, at pH 3.0 and pH 5.0 for Cu and Cd, respectively. The influence of the initial pH was ascertained (2.0 - 6.2 for Cu and 2.0 - 7.0 for Cd, respectively), at an initial metal ion concentration of 70 mg L^{-1} and a contact time of 960 minutes on the adsorption of Cu and Cd.

For competitive Cu, Cd and Zn adsorption experiments, a multi metal solution containing 0.5 mM of Cu, Cd and Zn (31.6, 56.2 and 32.6 mg L^{-1}), added as respective chloride salts, was prepared. 5 mL of the multi metal solution at different initial pH (2.5 - 6.1) was added to 5 mL or 10 mL of the BioSeNPs (0.29 g L^{-1}, pH 7.3). For competitive Cd and Zn adsorption experiments, 0.5 mM of Cd and Zn (32.6 and 56.2 mg L^{-1}), added as chloride salts, was prepared. 5 mL of this heavy metal solution at different initial pH (2.5 - 6.5) was added to 5 mL or 10 mL of BioSeNPs (0.29 g L^{-1}, pH 7.3). All the batch adsorption experiments were carried out under atmospheric conditions at 30 °C and 150 rpm in an orbital shaker.

It is important to note that for the adsorption experiments, the two solutions with different volume and different initial pH were mixed together, the concept of theoretical pH was used, as described in detail by (Jain et al., 2015). In short, the theoretical pH was obtained after calculating the final H^+ concentration in the samples after the addition of the metal ion solution and the BioSeNPs, without considering any adsorption reaction.

At the end of all adsorption experiments, the samples were filtered with a 0.45 µm syringe filter (cellulose acetate, Sigma Aldrich). Filtrates were analyzed for residual heavy metals and selenium concentrations using Atomic Adsorption Spectroscopy (AAS) and Inductively Coupled Plasma - Mass Spectrometry (ICP-MS), respectively. All experiments were carried out in duplicate and if the duplicate values varied by more than 10%, then the experiments were repeated. Average of the duplicate values and their errors are presented in the figures.

7.2.3. Characterization of heavy metals loaded on BioSeNPs

The ζ-potential of the BioSeNPs were measured with a Nano Zetasizer (Malvern instrument) at increasing heavy metals concentrations at the equilibrium pH of 5.4 as described in (Jain et al., 2015). The samples for SEM-EDXS and FT-IR were prepared after addition of 100 mg L^{-1} of Cu, Cd and Zn at initial metal ion pH of 3.0,

5.0 and 6.5, respectively, to 10 mL of 0.84 g L^{-1} and pH 7.3 BioSeNPs. For the multimetal FT-IR samples, equal metal concentrations in mg L^{-1} were added so that the total final metal concentration was 100 mg L^{-1}.

7.2.4. Analytics

SEM-EDXS, ICP-MS and AAS measurements were carried out according to the procedures described in (Jain et al., 2015). AAS for Cu and Cd were carried out using their respective lamps at 324.8 and 228.8 nm, respectively. For the FT-IR, the samples were vacuum dried and 1 mg of sample was measured. 300 mg of dried KBr was mixed with the samples, followed by pressing at 145,000 psi for 2 minutes to obtain clear KBr pellets. The spectroscopic data over the range 4000-400 cm^{-1} in the transmittance mode was recorded by a Bruker Vertex 70/v spectrometer equipped with a D-LaTGS-detector (L-alanine doped triglycine sulfate). Each spectrum was averaged out over 64 scans.

7.3. Results

7.3.1. SEM-EDXS analysis of BioSeNPs loaded with heavy metals

Figure 7.1a shows that the produced BioSeNPs are spherical in shape and composed of selenium (Figure 7.1b). EDXS analysis also suggests the presence of C, O, N and S on the BioSeNPs. These elements can be attributed to the presence of an EPS layer associated with the BioSeNPs (Jain et al., 2015). Figure 7.1c shows that there are no changes in the shape of the BioSeNPs after simultaneous loading with Cu, Zn and Cd. The presence of these heavy metals on the BioSeNPs was confirmed by EDXS analysis (Figure 7.1d). It is noteworthy to mention that the presence of Si in the EDXS analysis is due to the use of silicon wafer to hold the respective samples.

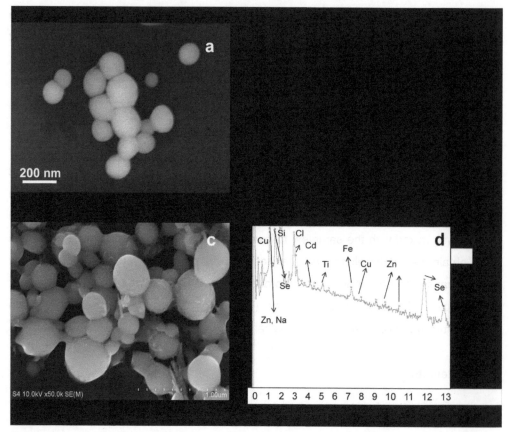

Figure 7.1. Secondary electron SEM image and EDXS analysis of (a), (b) BioSeNPs and (c), (d), BioSeNPs loaded with Cu, Zn and Cd. Figure a and b are reproduced from Jain et al. (2015).

7.3.2. Single component study

7.3.2.1. Adsorption time dependency

The equilibrium pH for the time-dependent adsorption experiments for Cu and Cd was 5.5 and 7.2, respectively. Time-dependent studies demonstrate that the adsorption of both copper and cadmium, like Zn as observed in a previous study (Jain et al., 2015) was fast and occurred mostly within the first 12 min (see inset) (Figures 7.2a, b and c). During the first five minutes of contact between the heavy metals and the BioSeNPs, nearly 84.3% of the maximum Qe-Cu (mg of Cu adsorbed per g of BioSeNPs) and 63.2% of maximum Qe-Cd (mg of Cd adsorbed per g of BioSeNPs) was adsorbed. 88% and 97% of the maximum Qe-Cu was achieved in 60 and 240 min of the reaction, respectively. In contrast, for cadmium, 85.7% and 93.2% of maximum Qe-Cd was achieved in 60 and 240 min, respectively. The maximum Qe-Cu and Qe-Cd was found to be 178.6 and 106.7, respectively after 220 min of contact time and remained constant afterwards. Hence, all the further adsorption experiments were carried out during 960 minutes to ensure that equilibrium is reached. It is important to note that Visual MINTEQ does not predict any precipitation of Cd at the conditions used in the time dependency study. However, Visual MINTEQ predicts 30% precipitation of Cu (tenorite) at the equilibrium pH 5.5 and equilibrium Cu concentrations of 30 mg L^{-1}. There was no evidence of Cu precipitation at the theoretical pH of 3.1 and concentrations of 70 mg L^{-1}.

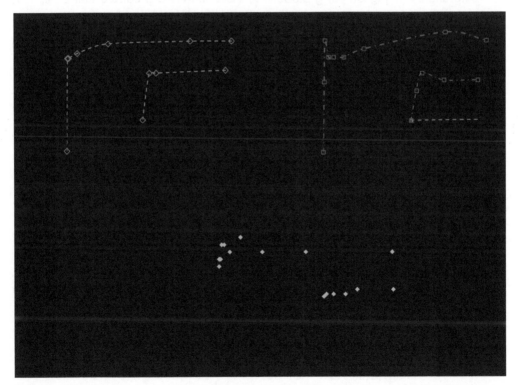

Figure 7.2. Single component batch adsorption experiments at the initial metal concentration of 70 mg L^{-1} with no background electrolyte with time for (a) Cu, (b) Cd and (c) Zn at equilibrium pH of 5.5, 7.2 and 6.5, respectively. Inset graphs are

zoomed in for the respective metal's adsorption kinetics. Figure 7.2c is reproduced from Jain et al. (2015).

7.3.2.2. Effect of pH

For the pH dependent Cu adsorption by BioSeNPs, the theoretical and equilibrium pH varied from 2.1 to 6.3 and 2.5 to 6.4, respectively (Figure 7.3a). The adsorption of Cu increased with an increase in theoretical and equilibrium metal ion pH (Figure 7.3a). The Qe-Cu increased to a value of 48.4 mg of Cu adsorbed per g of BioSeNPs when the theoretical and equilibrium pH values increased to 2.8 and 5.1, respectively. No Cu precipitation was predicted by Visual MINTEQ at these initial/theoretical (pH 3.0, 70 mg L^{-1}) and equilibrium conditions (pH 5.1, 56 mg L^{-1}) as noted by the black line in Figure 7.3a. When the Qe-Cu value reached to 114.3, 164.3 and 239.7 mg Cu per g BioSeNPs, Visual MINTEQ predicted that 23%, 60% and 95% of the Cu, respectively, was precipitated, mainly as tenorite. Like Qe-Cu, the Qe-Cd also increased with an increase in the theoretical and equilibrium pH. The maximum Qe-Cd of 114.3 mg of Cd per g of BioSeNPs was obtained when the theoretical and equilibrium pH values were 7.1 and 7.4, respectively (Figure 7.3b). There was no precipitation predicted by Visual MINTEQ at all the theoretical and equilibrium conditions. The effect of pH on adsorption of Zn onto BioSeNPs has been described in a previous study (Jain et al., 2015). It is important to note that the metals to BioSeNPs ratio (v/v) is identical for Cu and Cd, but not same for Zn. Thus, the absolute Qe-metals values cannot be compared among different heavy metals from Figure 7.3.

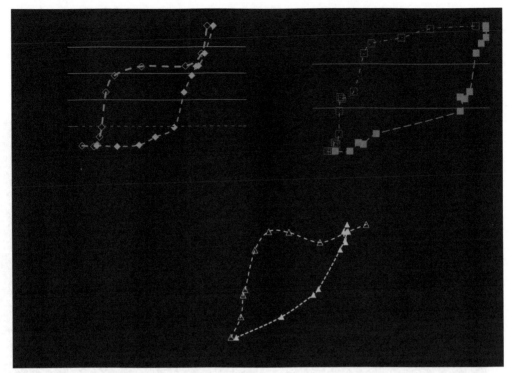

Figure 7.3. Single component adsorption of (a) Cu, (b) Cd and (c) Zn onto BioSeNPs with the change in theoretical (Cu, Cd and Zn : ◊, □ and △, respectively) and equilibrium pH (Cu, Cd and Zn : ♦, ■ and ▲, respectively) at an initial metal concentration of 70 mg L^{-1} and no background electrolyte. The zone below the black dash line in Figure 3a represents no precipitation zone. Please note that Zn adsorption data have been reproduced from (Jain et al., 2015).

7.3.3. Multimetal batch adsorption experiments

7.3.3.1. Cu, Cd and Zn

The objective of this experiment was to assess if the pH and metal to BioSeNPs ratio can be varied to optimize the preferential adsorption of one of the supplied metals (Cu, Cd and Zn). Simulations by Visual MINTEQ suggest that there was no precipitation at the theoretical and equilibrium conditions (respective, pH values and metal concentrations). When the theoretical pH was increased from 3.0 to 4.5 and 2.8 to 4.8 for metals to BioSeNPs ratios (v:v) of 1:2 and 1:1, respectively, the removal of the copper ion increased to 79.8 and 45.2% for 1:2 and 1:1 ratios, respectively (Figure 7.4a). The equilibrium pH increased from 3.6 to 5.6 and 3.1 to 5.3 during this theoretical pH increase for the 1:2 and 1:1 ratios, respectively (Figure 7.4b).

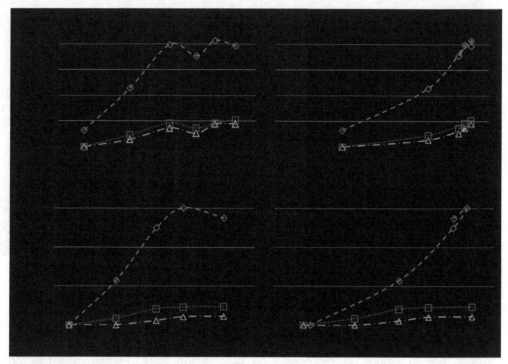

Figure 7.4. Removal of Cu (◊), Cd (□) and Zn (Δ) at different metals to BioSeNPs ratios with change in (a) theoretical pH and (b) equilibrium pH at ratio of 1:2 and with change in (c) theoretical pH and (d) equilibrium pH at ratio of 1:1.

A further increase in theoretical pH led to less than 10% and 5% increase in metal removal and equilibrium pH for both ratios. For ratios 1:2 and 1:1, the maximum percentage of copper removed was 83.0% and 61.6%, respectively. The maximum percentage of cadmium removed for the ratio 1:2 and 1:1 was 21.1% and 8.6%. The maximum percentage of zinc removed for the ratio 1:2 and 1:1 was 17.8% and 3.2%, respectively.

7.3.3.2. Cd and Zn

As Cu was shown to be selectively adsorbed from the metal solution containing Cu, Cd and Zn, it was important to elucidate if the pH or metal to BioSeNPs ratio can be further optimized to preferentially adsorb either Cd or Zn. To this point, the metal solution containing Cd and Zn was contacted with to BioSeNPs at different theoretical pH and metal to BioSeNPs ratios.

The increase in the theoretical and equilibrium pH of the samples with a metals to BioSeNPs ratio of 1:2 (v:v) show two different slopes for both Cd and Zn removal (Figure 7.5a, b). At the theoretical and equilibrium pH of 3.8 and 6.5, respectively, the end of first slope was observed with a Zn and Cd removal efficiency of 66.1% and 58.2%, respectively. The end of the second slope was observed when the theoretical and equilibrium pH of 7.0 was obtained, where the maximum Zn (80.5%) and Cd (76.3%) removal efficiency was achieved for metals to BioSeNPs ratio of 1:2 (v:v). Interestingly, no selective adsorption of either Zn or Cd was observed at this metals to BioSeNPs ratio.

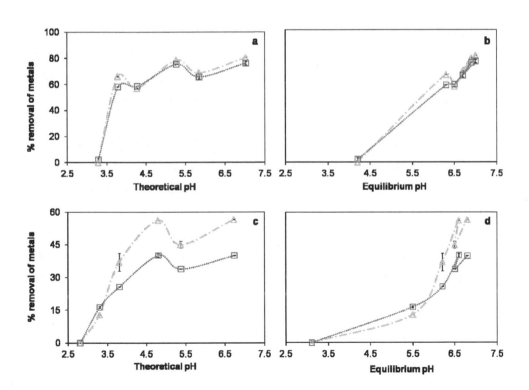

Figure 7.5. Removal of Cd (□) and Zn (△) at different metals to BioSeNPs ratios with change in (a) theoretical pH and (b) equilibrium pH at ratio of 1:2 and with change in (c) theoretical pH and (d) equilibrium pH at ratio of 1:1.

At the metals to BioSeNPs ratio of 1:1 (v/v), no two distinct slopes for Zn or Cd removal was observed (Figures 7.5c, d). Both Zn and Cd removal increased with an increase in theoretical and equilibrium pH. The maximum Zn (56.6%) and Cd (40.0%) removal efficiency was achieved at theoretical and equilibrium pH of 6.7 and 6.8, respectively. At this metal to BioSeNPs ratio, Zn was 16% more adsorbed than Cd suggesting a slight preference of BioSeNPs towards Zn. It should be noted that simulations by Visual MINTEQ did not indicate any precipitation of Cd and Zn at these theoretical and equilibrium conditions (respective, pH and metals concentrations).

7.3.4. BioSeNPs flow-through in single, binary and ternary experiments

BioSeNPs flow-through in the filtrate or retention in the filter was measured for single and multimetal adsorption experiments. For pH dependent single metal ion batch adsorption experiments of Cu, the flow-through of BioSeNPs dropped from 920 to 265 µg L^{-1} with an increase in the theoretical and equilibrium pH values to 3.1 and 5.7, respectively (Figures 7.6a, b). With a further increase in the theoretical and equilibrium pH from 3.3 to 6.3 and 5.7 to 6.4, respectively, there was a slight drop in selenium concentration in the filtrate, from 265 to 243 µg L^{-1}. The residual selenium concentration or flow-through BioSeNPs in the filtrate without adsorption was 13710 µg L^{-1}, which was 56.4 times higher than the lowest selenium concentration observed in the filtrate after adsorption of copper ions. It is important to note that the selenium concentration in the filtrate was 23.1 times lower when no precipitation of Cu was observed compared to the selenium concentration or flow-through BioSeNPs after filtration, but without adsorption.

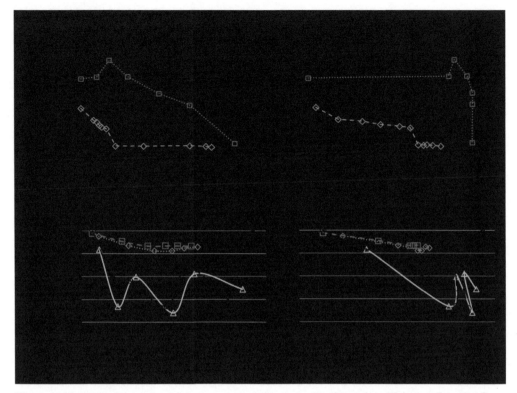

Figure 7.6. Variations of selenium concentrations in the filtrate or flow-through of the single metal adsorption experiments of Cu (◊) and Cd (□) with change in (a) theoretical pH and (b) equilibrium pH. Percentage removal of added BioSeNPs in the filtrate during the simultaneous Cu, Cd and Zn adsorption experiment (◊ : metals to BioSeNPs ratio of 1:2; □ : metals to BioSeNPs ratio of 1:1) and simultaneous Cd and Zn adsorption experiments (○ : metals to BioSeNPs ratio of 1:2; △ : metals to BioSeNPs ratio of 1:2) adsorption experiments with (c) theoretical and (d) equilibrium pH.

As observed previously in the pH dependent adsorption study of Cu, the residual selenium concentration or flow-through of BioSeNPs decreased as the pH increased in the adsorption study of Cd (Figure 7.6b). However, unlike observed in the pH dependent adsorption study of Cu, the residual selenium concentration or flow-through of BioSeNPs for the cadmium experiments did not drop sharply and remained close to 1434 µg L^{-1} when the theoretical and equilibrium pH was increased from 2.2 to 3.7 and from 2.3 to 7.2. When the theoretical pH value was increased further from 3.7 to 7.1 (equilibrium pH changed from 7.2 to 7.4), the residual selenium concentration or flow-through of BioSeNPs dropped from 1434 to 324 µg L^{-1}. The residual selenium concentration or flow-through BioSeNPs in the filtrate after filtration but without adsorption was 42.3 times higher than the lowest selenium concentration observed in the filtrate after cadmium adsorption.

The removal of BioSeNPs in the multi-component experiments was calculated by comparing with the selenium concentration obtained when the BioSeNPs were mixed with Milli-Q water (at the same metals to BioSeNPs ratio: 1:1 and 1:2) and filtered but without any adsorption of heavy metals. For the Cu, Cd and Zn experimental system, selenium removal decreased from 98.8% to 93.3% when the theoretical and equilibrium pH increased from 2.8 to 5.5 and 3.1 to 5.4, respectively, at the metals to BioSeNPs ratio of 1:1. When the theoretical and equilibrium pH was increased from 3.0 to 5.6 and 3.6 to 5.8, the selenium concentration in the filtrate decreased from 97.5 to 92.9% when the metals to BioSeNPs ratio was 1:2 (Figures 7.6c, d).

For the Cd and Zn adsorption experiments, when the theoretical and equilibrium pH increased from 2.8 to 6.7 and 3.1 to 6.8, respectively, the selenium concentrations in the filtrate decreased from 100.0 to 66.0% when the metals to BioSeNPs ratio was 1:1. The selenium concentration decreased from 91.7% to 74.4% when the theoretical and equilibrium pH increased from 3.0 to 6.8 and 4.2 to 7.0, respectively, when the metals to BioSeNPs ratio was 1:2.

7.3.5. Effect of heavy metals loading on the ζ-potential of BioSeNPs

The ζ-potential of BioSeNPs loaded with Cu and Cd ions became less negative when the initial metal concentrations were increased (Figure 7.7). The equilibrium pH varied from 5.5 to 6.2 in all cases investigated. Similar observations were made with BioSeNPs loaded with Zn (Jain et al., 2015) which are being reproduced in Figure 7.7. The ζ-potential of BioSeNPs loaded with copper ions was more negative than BioSeNPs loaded with cadmium or zinc ions when the initial metal ion concentration exceeded 40 mg L^{-1} (Figure 7.7). It is important to note that the change in the ζ-potential of the BioSeNPs without being exposed to heavy metals at the equilibrium pH of 5.5 to 6.2 is less than -5 mV as observed in earlier studies (Jain et al., 2015).

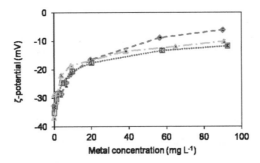

Figure 7.7. ζ-potential variations of BioSeNPs loaded with Cu (◊), Cd (□) and Zn (△) (a) at an increasing metal ions concentrations. Please note that Zn adsorption data are from Jain et al. (2015).

7.3.6. FT-IR analysis of BioSeNPs prior and after loading with heavy metals

The FT-IR analysis of BioSeNPs has been described in detail in a previous study (Jain et al., 2014). Briefly, FT-IR analysis revealed the presence of hydroxyl (3435 cm^{-1}), hydrocarbons (2924 - 2855 cm^{-1}), proteins (amide I - 1637 cm^{-1}, amide II - 1533 cm^{-1} and amide III - 1239 cm^{-1}), methyl groups (1446 cm^{-1}), carboxyl groups (1403 and 1388 cm^{-1}) and carbohydrates (1075 to 1032 cm^{-1}) as described in (Wang et al., 2012; Xu et al., 2011).

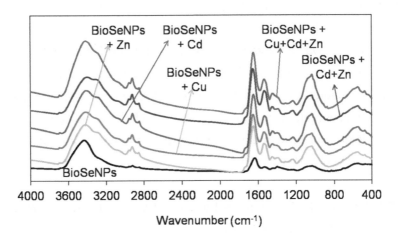

Figure 7.8. FTIR spectra of BioSeNPs (—) and BioSeNPs loaded with Zn (—), Cu (—), Cd (—), Zn+Cd (—) and Cu+Cd+Zn (—).

BioSeNPs loaded with the heavy metals in single (Cu, Cd, and Zn) and multimetal (Cd+Zn, Cu+Cd+Zn) systems showed subtle differences as compared to BioSeNPs that were not exposed to heavy metals. There was a distinct rise of the feature at 3290 cm^{-1} in all the BioSeNPs loaded with heavy metals, while this feature was not present in the BioSeNPs not exposed to heavy metals. The strong feature at 1403 cm^{-1} present in the BioSeNPs not exposed to heavy metals disappeared in the BioSeNPs loaded with heavy metals. There was also a rise of the feature at 1313 cm^{-1}, most likely corresponding to S-O stretching, in the BioSeNPs exposed to heavy metals which was absent in the BioSeNPs not exposed to heavy metals.

7.4. Discussion

7.4.1 Optimized pH and metals to BioSeNPs ratio for selective adsorption of heavy metals

This study demonstrated that the pH and metal to BioSeNPs ratio can be used to selectively adsorb Cu from an equimolar Cu, Zn and Cd mixture. The ratio of the percentage of Cu removal to the sum of Zn and Cd removal was used to observe the optimized pH and metal to BioSeNPs ratio for the maximum preferential adsorption of Cu onto BioSeNPs. Table 1 shows that the maximum ratio observed for the metal to BioSeNPs ratio of 1:1 and 1:2 were 6.6 and 3.1, respectively. However, at the ratio of 6.6, only 17.2% of total Cu was adsorbed, the next best ratio was 4.7. The ratio of 4.7 and 3.1 was achieved at a theoretical pH of 4.3 and 3.8 for the metal to BioSeNPs ratio of 1:1 and 1:2, respectively. Interestingly, the percentage of Cu removal at this theoretical pH was 45.2 and 45.8% for the metal to BioSeNPs ratio of 1:1 and 1:2, respectively. This suggests that for preferential adsorption, the influent pH of the metal solution or metals to BioSeNPs ratio could be modified to futher optimize the preferential adsorption.

Table 7.1. Ratio of % of Cu removed to sum of % of Cd and Zn removed in the multimetal (Cu+Cd+Zn) system.

| Metals to BioSeNPs ratio | | | | | |
| 1:1 | | | 1:2 | | |
Theoretical pH	Equilibrium pH	Ratio	Theoretical pH	Equilibrium pH	Ratio
2.8	3.1	-	3.0	3.6	-
3.6	4.5	6.6	3.8	5.0	3.1
4.3	5.3	4.7	4.5	5.6	2.4
4.8	5.5	4.5	4.9	5.5	2.8
5.5	5.4	4.0	5.3	5.7	2.3
-	-	-	5.6	5.7	2.0

For the Cd and Zn containing metal solution, the ratio of % removal of Zn to Cd was close to 1 when the metal to BioSeNPs ratio was 1:2. The maximum ratio achieved was 1.4 above the theoretical pH of 3.8 for metals to BioSeNPs ratio of 1:1. In the binary system (Zn and Cd containing metal solution), the theoretical pH did not affect the preferential adsorption of Zn or Cd, while the metal to BioSeNPs ratio (v/v) had a limited effect on the preferential adsorption of either Zn or Cd.

For both multi-metal systems investigate (Cu, Cd, Zn and Cu, Cd), the metal to BioSeNPs ratio of 1:1 is more preferred for preferential adsorption. This may be due to the fact that lower concentrations of BioSeNPs limit the number of sites, resulting in a stronger competition for the existing sites and hence more preferential adsorption.

7.4.2. Selective adsorption of heavy metals by BioSeNPs

This study demonstrated that BioSeNPs showed preference for the metals adsorbed in the order of Cu>Zn>Cd. This fact was further confirmed by the lower ζ-potential of Cu loaded BioSeNPs as compared to Cd and Zn loaded BioSeNPs at equal metal concentrations (Figure 7). Selective adsorption of Cu among the other heavy metals present in the mixture was also observed in adsorption studies with untreated coffee husks (Oliveira et al., 2008), graphene oxide nanoparticles (Sitko et al., 2013) and marine algal biomass (Sheng et al., 2004).

The preferential adsorption of heavy metals onto any adsorbent occurs following the lower ionic radius, the higher electronegativity, or the higher ratio of ionization potential to ionic radius (McKay and Porter, 1997). Among the metals investigated in this study, i.e. Cu, Cd and Zn, Cu has the lowest ionic radius and highest electronegativity and ratio of ionic potential to ionic radius (Table 2). Metal ions are also known to hydrolyze or to complex with acetate or equivalent carboxylic functional groups present on the surface of adsorbent (Sitko et al., 2013). So, the relative ease of the formation of metal hydroxo species or acetate complexes also suggests the relative preference of the metals towards the adsorbent. The first stability constant of metal hydroxo and acetate formation is also higher for Cu, followed by Zn and Cd (Table 2) (Dean, 1999).

Table 7.2. Chemical properties of Cu, Cd and Zn.

Properties	Cu^{2+}	Cd^{2+}	Zn^{2+}
Ionic radius (A°)	0.73	0.95	0.74
Pauling electronegativity (a.u.)	1.90	1.69	1.65
*Std. reduction potential vs. NHE(V) $M^{2+} + 2e^- = M$	0.34	-0.40	-0.76
$\log K_1$ - $Me^{2+} + OH^- = Me(OH)^+$	7.0	4.177	4.4
$\log K_1$ - $Me^{2+} + Ac^- = Me(Ac)^+$	2.16	1.5	1.5

*Note: NHE refers to normal hydrogen electrode

The formation of metal hydroxo and metal acetates complexes in this study was evident from the changes in the FT-IR spectra of BioSeNPs compared to the BioSeNPs loaded with heavy metals (Figure 8). The rise of the feature at 3290 cm^{-1} suggests the interaction of the hydroxyl and metal ions. Though a difference in the water contents could cause this shift, the presence of this shift in all the BioSeNPs loaded with metal ions rules out the possibility of this artifact. Similarly, the disappearance or shift of the feature at 1403 cm^{-1} observed in the BioSeNPs, but absent in all the samples of BioSeNPs loaded with heavy metals, suggests the interaction of heavy metals with the carboxyl functional groups.

Zn is 16% more preferentially adsorbed onto BioSeNPs when Zn and Cd are present at a metal to BioSeNPs ratio of 1:1 (Figure 5c). While when the metal to BioSeNPs ratio was 1:2, there was no preferential adsorption of Zn. The ionic radius of Zn is smaller than Cd and their electronegativity is similar, but the ratio of ionization potential to ionic radius is lower (−0.42 for Cu as compared to −1.02 for Zn) (Table 2). Thus, there should not be much preference towards either Zn or Cd. This is also evident from various studies where in some cases Zn is preferentially adsorbed onto *Sphaerotilus natans* (Pagnanelli et al., 2003) and in some cases Cd is preferentially adsorbed on graphene oxide (Sitko et al., 2013) or marine algal biomass (Sheng et al., 2004).

7.4.3. Colloidal stability of BioSeNPs

The colloidal stability of the BioSeNPs decreases with an increase in the metal loading and decrease in the pH as observed in previous studies (Buchs et al., 2013; Jain et al., 2015; Staicu et al., 2014). This study also showed that the concentration of selenium in the filtrate decreases (colloidal stability decreases) with an increase in the heavy metals owning to a larger loading of Cu and Cd onto BioseNPs with increasing pH (Figures 7.3, 7.6a, b). The increased loading of Cu and Cd neutralizes the negative charge on the surface of BioSeNPs leading to a lower colloidal stability and lower selenium concentration in the filtrate. The neutralization of the negative charge on the surface of BioSeNPs was done more effectively by Cu compared to Cd, asevidenced by the lower selenium concentration in the filtrate after adsorption and filtration (Figures 7.6a, b). This observation is also evident from Figure 7.7 where copper loaded BioSeNPs have lower ζ-potential wheexposing BioSeNPs to an equal individual metal concentration. The less negative ζ-potential of BioSeNPs loaded with Cu as compare to BioSeNPs loaded with Cd or BioSeNPs loaded with Zn is due to a higher preference of Cu to the properties or the functional groups of the BioSeNPs surface. This finding is further evident when the BioSeNPs loaded with Cu, Zn and Cd were effectively retained in the filter or removed in the filtrate (Figures 7.6c, d). However, when the BioSeNPs were loaded with only Cd and Zn, the retention of the BioSeNPs was almost 10% less than at near-neutral pH values when compared to the Cu, Cd and Zn system. This further demonstrates the affinity of Cu to the BioSeNPs and the effect of Cu in decreasing the colloidal stability of BioSeNPs and hence improving their retention in the filter or removal from the filtrate.

7.4.4. Effect of pH on the adsorption of heavy metals

The increase in the initial metal solution and equilibrium pH leads to an increase in the Qe-Metal for both Cu and Cd. The increase in the pH leads to deprotonation of the sites of the BioSeNPs that are made available for the metals ions to interact (Arias and Sen, 2009). This deprotonation of the sites with increase in pH also makes the ζ-potential of the BioSeNPs less negative, thus increasing the electrostatic attraction between BioSeNPs and metal ions, and hence leading to a higher adsorption. The increase in the pH also leads to precipitation, especially in the case of Cu leading to high Qe-Cu values. Indeed, precipitation was predicted by Visual MINTEQ and also observed in the Figures 2 and 3a when the initial and equilibrium pH exceeded 5.5. Since, the equilibrium pH of the time-dependency experiments was close to 5.5, the presence of precipitation cannot be excluded. The highest Qe-Cu values observed without any sign of precipitation was 48.4 mg of Cu adsorbed per g of BioSeNPs (Figure 3). The maximum Qe-Cd values observed was almost 2 times higher than the one observed for Cu. This suggest that at higher pH, adsorption of Cu onto BioSeNPs is competing with precipitation of Cu mainly as tenurite, as predicted by Visual MINTEQ. The Qe-Zn for zinc adsorption onto BioSeNPs observed in the earlier study (Jain et al., 2015) was 48% lower than the ones observed in this study for Cu and Cd. This might be due to the different heavy metals to BioSeNPs ratios used in these studies.

7.5. Conclusions

This study demonstrates that Qe-Metal for Cu and Cd is 48.4 and 114.3 mg of heavy metal per g of BioSeNPs without the occurrence of precipitation. The higher Qe-Cu values of 239.7 mg of Cu adsorbed per g of BioSeNPs observed was most likely due to precipitation. In the multimetals adsorption experiments, BioSeNPs adsorbed 4.7 times more Cu than combined Zn and Cd at the theoretical pH of 4.3 when the metal to BioSeNPs ratio was 1:1. This preference towards Cu can be attributed to the chemical properties of Cu and functional groups present on the BioSeNPs. FT-IR data suggest that BioSeNPs might be interacting through hydroxyl and carboxyl groups to the heavy metals. The retention of BioSeNPs on the filter increased with an increase in the pH when BioSeNPs were loaded with heavy metals. This study also demonstrated that Cu is effective in neutralizing the surface charge on BioSeNPs and leading to their removal from the filtrate.

Acknowledgements

The authors would like to acknowledge Ferdi Battles and Berend Lolkema (UNESCO-IHE, Delft) for adsorption experiments and ICP-MS analysis.

7.6 References

Arias, F., Sen, T.K., 2009. Removal of zinc metal ion (Zn2+) from its aqueous solution by kaolin clay mineral: A kinetic and equilibrium study. Colloids Surfaces A Physicochem. Eng. Asp. 348, 100–108.

Buchs, B., Evangelou, M.W.-H., Winkel, L., Lenz, M., 2013. Colloidal properties of nanoparticular biogenic selenium govern environmental fate and bioremediation effectiveness. Environ. Sci. Technol. 47, 2401–2407.

Dean, J.A., 1999. Lange's handbook of chemistry, Fifeteenth. ed. McGraw-Hill,, New York.

Fu, F., Wang, Q., 2011. Removal of heavy metal ions from wastewaters: a review. J. Environ. Manage. 92, 407–418.

Hua, M., Zhang, S., Pan, B., Zhang, W., Lv, L., Zhang, Q., 2012. Heavy metal removal from water/wastewater by nanosized metal oxides: a review. J. Hazard. Mater. 211-212, 317–331.

Jain, R., Jordan, N., Schild, D., Hullebusch, E.D. Van, Weiss, S., Franzen, C., Hubner, R., Farges, F., Lens, P.N.L., 2015. Adsorption of zinc by biogenic elemental selenium nanoparticles. Chem. Eng. J. 260, 850–863.

Jain, R., Jordan, N., Weiss, S., Foerstendorf, H., Heim, K., Kacker, R., Hübner, R., Kramer, H., van Hullebusch, E.D., Farges, F., Lens, P.N.L., 2014. Extrapolymeric celluar substances govern the surface charge of biogenic elemental selenium nanoparticles. Environ. Sci. Technol. *(Submitted, Chapter 3)*

Kanawade, S.M., Gaikwad, R.W., 2011. Removal of zinc ions from industrial effluent by using cork powder as adsorbent. Int. J. Chem. Eng. Appl. 2, 199–201.

Lenz, M., Smit, M., Binder, P., van Aelst, A.C., Lens, P.N.L., 2008. Biological alkylation and colloid formation of selenium in methanogenic UASB reactors. J. Environ. Qual. 37, 1691–700.

Lenz, M., Van Aelst, A.C., Smit, M., Corvini, P.F.X., Lens, P.N.L., 2009. Biological production of selenium nanoparticles from waste waters. Mater. Res. 73, 721–724.

McKay, G., Porter, J.F., 1997. Equilibrium Parameters for the Sorption of Copper, Cadmium and Zinc Ions onto Peat. J. Chem. Technol. Biotechnol. 69, 309–320.

Oliveira, W.E., Franca, A.S., Oliveira, L.S., Rocha, S.D., 2008. Untreated coffee husks as biosorbents for the removal of heavy metals from aqueous solutions. J. Hazard. Mater. 152, 1073–1081.

Pagnanelli, F., Esposito, a, Toro, L., Vegliò, F., 2003. Metal speciation and pH effect on Pb, Cu, Zn and Cd biosorption onto Sphaerotilus natans: Langmuir-type empirical model. Water Res. 37, 627–633.

Sheng, P.X., Ting, Y.-P., Chen, J.P., Hong, L., 2004. Sorption of lead, copper, cadmium, zinc, and nickel by marine algal biomass: characterization of biosorptive capacity and investigation of mechanisms. J. Colloid Interface Sci. 275, 131–141.

Sitko, R., Turek, E., Zawisza, B., Malicka, E., Talik, E., Heimann, J., Gagor, A., Feist, B., Wrzalik, R., 2013. Adsorption of divalent metal ions from aqueous solutions using graphene oxide. Dalton Trans. 42, 5682–5689.

Staicu, L.C., van Hullebusch, E.D., Lens, P.N.L., Pilon-Smits, E.A., Oturan, M. a, 2014. Electrocoagulation of colloidal biogenic selenium. Environ. Sci. Pollut. Res. Int. (accepted) doi:10.1007/s11356-014-3592-2.

United States Environmental Protection Agency website, 2013. National recommended water quality criteria. http://water.epa.gov/drink/contaminants/secondarystandards.cfm. (access date: 14/09/2014).

Wang, L.-L., Wang, L.-F., Ren, X.-M., Ye, X.-D., Li, W.-W., Yuan, S.-J., Sun, M., Sheng, G.-P., Yu, H.-Q., Wang, X.-K., 2012. pH dependence of structure and surface properties of microbial EPS. Environ. Sci. Technol. 46, 737–744.

Wingenfelder, U., Hansen, C., Furrer, G., Schulin, R., 2005. Removal of heavy metals from mine waters by natural zeolites. Environ. Sci. Technol. 39, 4606–4613.

Xu, C., Zhang, S., Chuang, C., Miller, E.J., Schwehr, K. a., Santschi, P.H., 2011. Chemical composition and relative hydrophobicity of microbial exopolymeric substances (EPS) isolated by anion exchange chromatography and their actinide-binding affinities. Mar. Chem. 126, 27–36.

Yan, H., Dai, J., Yang, Z., Yang, H., Cheng, R., 2011. Enhanced and selective adsorption of copper (II) ions on surface carboxymethylated chitosan hydrogel beads. Chem. Eng. J. 174, 586–594.

CHAPTER 8

Selenate removal from wastewater by mesophilic and thermophilic UASB reactors

This chapter will be published as

Dessi, P., Jain, R., Singh, S., Seder-Colomina, M., Kacker, R., Kramer, H., Hullebusch, E.D. Van, Farges, F., Lens, P.N.L., 2015. Selenate removal from wastewater by mesophilic and thermophilic UASB reactors *(in preparation)*.

Abstract:

The effect of temperature and nitrate on the biological anaerobic reduction of selenate in an upflow anaerobic sludge blanket (UASB) reactor was investigated under mesophilic (30 $^\circ$C) and thermophilic (55 $^\circ$C) conditions. Under both conditions, selenate was effectively removed from 790 µg L^{-1} (10 µM) to less than 50 µg L^{-1} in the effluent in a UASB reactor (pH 7 + hydraulic retention time 8 h) operating with lactate as the electron donor at an organic loading rate of 0.5 gCOD L^{-1}.d^{-1}. At a feed concentration of 50 µM of selenate (3950 µg L^{-1}), the thermophilic reactor achieved a higher total selenium removal efficiency than the mesophilic reactor (~ 93% and ~ 85%, respectively). The dissolved selenium removal efficiencies were, however, similar, i.e. 95.2% and 94.9% for the thermophilic and for mesophilic UASB reactors, respectively, suggesting the presence of elemental selenium in the effluent. When 100 and 500 µM nitrate was added to the synthetic wastewater, the selenate removal efficiency under mesophilic conditions was not affected. In contrast, 100 µM nitrate initially affected the reduction of selenate under thermophilic conditions (total selenium removal efficiency dropped to 85.3%), but the total selenium removal efficiency increased to 92.2% when the nitrate concentration was increased to 500 µM.

Keywords: Selenate reduction, thermophilic, nitrate, dissolved selenium, UASB reactor

Graphical abstract:

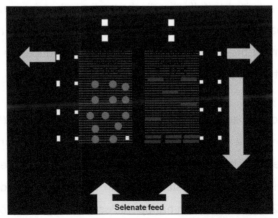

149

8.1. Introduction

Selenium is considered as an "essential toxin" for humans (Lenz and Lens, 2009) because there is only one order of magnitude difference between the nutritious requirement (30-85 µg Se d^{-1}) and toxicity (400 µg Se d^{-1}) level. The deleterious effect of selenium is not due only to its concentration, but also due to its speciation in the environment. In fact, both volatile and dissolved forms of selenium are more toxic compared to solid elemental selenium.

Due to the toxicity of selenium, the United States Environmental Protection Agency (US EPA) has set a treatment standard at 50 µg Se L^{-1} and the recommended limit is 5 µg Se L^{-1} (USEPA, 2001). Both physical (nanofiltration, reverse osmosis) and chemical (ion exchange, ferrihydrite absorption, zero valent iron absorption) methods have been widely tested for selenium removal from wastewaters (Frankenberger et al., 2004; NSMP, 2007; USEPA, 2001; Zhang et al., 2005). However, their application in full-scale systems is rather limited due to low efficiencies and economic reasons. Biological remediation including microbial reduction of selenium oxyanions into elemental selenium in bioreactors is considered a promising option (Lenz et al., 2009; Lenz et al., 2008; Soda et al., 2011). Indeed, pilot scale bioreactors have been demonstrated to reduce selenate from agriculture drainage wastewater (Cantafio et al., 1996). However, the biological reduction of selenium oxyanions always results in the production of colloidal elemental selenium nanoparticles that are present in the effluent of the reactor (Buchs et al., 2013; Jain et al., 2015). If the concentration of nanoparticles in the effluent is too high, a second treatment step is required to fulfil the discharge limit, thus increasing the operating costs. Another challenge is the presence of competitive electron acceptors such as nitrate that can hinder the selenate reduction and thus reduce the selenium removal efficiency (Lenz et al., 2009).

Biological reduction of selenate under thermophilic conditions (55°C) can be an option to overcome the challenges that are commonly encountered during the treatment of selenium laden wastewater. It has been reported that a rise in temperature can increase the crystallinity (Jain et al., 2014) and size (Lee et al., 2007; Tam et al., 2010) of selenium nanoparticles, allowing them to settle without the addition of any coagulant. It is also known that the bacterial populations change significantly with a temperature transition from mesophilic to thermophilic conditions (Khemkhao et al., 2012; Li et al., 2014). In comparison to mesophilic conditions, under thermophilic conditions microorganisms could evolve in a such a way that their selenate removal efficiency is not affected by the presence of nitrate.

In this study, the biological reduction of selenate at thermophilic (55°C) conditions was investigated in an UASB reactor inoculated with anaerobic granular sludge. A second reactor, operating at identical conditions but at mesophilic (30°C) conditions was used as a control. The research focused on the effect of temperature on the total and dissolved selenium removal efficiency and selenium speciation in the effluent. Both reactors were fed with a synthetic wastewater (pH 7.0) containing selenate, lactate as electron donor, micro and macro nutrients and trace metals. The effect of nitrate at different concentrations (100 µM, 500 µM and 5000 µM) on the total and dissolved selenium removal efficiency in both reactors was also investigated. Denaturated Gradient Gel Electophoresis was carried out to quantify changes in the microbial communities of the UASB reactors.

8.2. Materials and methods

8.2.1. Source of biomass

The seed sludge originated from a full scale UASB reactor treating wastewater of four paper mills (Industriewater Eerbek B.V., Eerbek, The Netherlands) and has been described in detail by Roest et al. (2005). Both reactors were inoculated with anaerobic granular sludge, 200 g wet weight, as described by Lenz et al. (2008).

8.2.2. Composition of the synthetic wastewater

The reactors were fed with a synthetic selenate containing wastewater. The composition of the synthetic wastewater as well as the acid and alkaline trace metal solution is summarized in Tables 8.1, 8.2 and 8.3, respectively. Lactate was used as sole electron donor at an organic loading rate of 0.5 gCOD.L^{-1}.d^{-1}. Sodium selenate

(Na_2SeO_4) and potassium nitrate (KNO_3) were added at different concentrations during the experiment, as explained in the section 8.24.

Table 8.1. Composition of synthetic wastewater

Compound	Concentration (g L^{-1})
$Na_2HPO_4 \cdot 2H_2O$	0.053
KH_2PO_4	0.041
NH_4Cl	0.300
$CaCl_2 \cdot 2H_2O$	0.010
$MgCl_2 \cdot 6H_2O$	0.010
$NaHCO_3$	0.040
Acid trace metals solution*	0.100
Alkaline trace metals solution*	0.100

*Acid and alkaline trace metals solution was added in mL L^{-1}

Table 8.2. Composition of acid trace metals solution

Compound	Concentration (mM)
$FeCl_2$	7.5
H_3BO_4	1
$ZnCl_2$	0.5
$CuCl_2$	0.1
$MnCl_2$	0.5
$CoCl_2$	0.5
$NiCl_2$	0.1
HCl	50

Table 8.3. Composition of alkaline trace metals solution

Compound	Concentration (mM)
Na_2WO_4	0.1

Na$_2$MoO$_4$	0.1
NaOH	10

8.2.3. UASB setup and operating conditions

Two identical UASB reactors were used during the experiments (Figure 8.1). The operating parameters are summarized in the Table 8.4. One of the reactors was maintained at 30°C, while the other one was maintained at 55°C with the help of a water jacket. The scheme of the reactors is shown in Figure 8.1c. The sampling ports of both reactors were called S1, S2 and S3 from the bottom to the top, respectively. Please note that the influent was fed from the bottom and the recirculation ratio of 2 was maintained for effective mixing. Samples were collected daily from the influent and effluent. The samples from the different sampling ports for profiling of elemental selenium in the reactor were taken in period II. Gas samples were analyzed at the end of every period. Volatilized selenium was trapped by collecting vapour phase in two gas traps. The first one (G1) contained 200 ml of concentrated HNO$_3$ (12 M) with the purpose to trap alkylated selenium compounds (DMSe and DMDSe); the second one (G2) contained 100 ml of 6M NaOH was used to trap H$_2$Se.

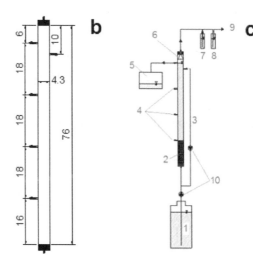

Figure 8.1. (a) Photograph of the UASB reactors used in this study; (b) Dimensions in cm of the reactors and (c) Schematic overview of the reactors. Influent tank (1), anaerobic sludge (2), recirculation system (3), sampling ports S1, S2 and S3 from the bottom to the top (4), effluent tank (5), gas seperator (6), HNO_3 trap (7), NaOH trap (8), gas outlet (9) and peristaltic pumps (10).

Table 8.4. Operating parameters of the reactors

Volume (ml)	HRT (h)	Influent flow (ml min^{-1})	Recirculation ratio	Recirculation flow (ml min^{-1})	Upflow velocity (cm h^{-1})
1000	8	2.2	2	4.4	27.3

8.2.4. Experimental conditions

Table 8.5 summarizes the operating conditions for both the reactors. Both reactors were operated in excess of electron donor for the entire duration of the study. Lactate concentration was increased 10 times on day 85. During period I, sodium selenate (Na_2SeO_4) was added to the influent solution at a concentration of 10 µM, as operated in the study carried out by Lenz et al. (2008a). The pH was adjusted to 7.0 - 7.5. During period II, the biomass was removed and the UASB reactors were restarted keeping the previous conditions, but increasing the selenate concentration to 50 µM. During period III, nitrate (100 µM) was added to the influent along with selenate (50 µM). The concentration of nitrate was increased 5 times (500 µM) in period IV and 10 times more (5000 µM) during period V.

Table 8.5. Operating conditions of UASB reactors during different operational periods.

Period	Days	SeO$_4^{2-}$ (µM)	NO$_3^-$ (µM)	Lactate (µM)
I	1-25	10	0	1.38
II[a]	26-43	50	0	1.38
III	44-60	50	100	1.38

| IV | 61-82 | 50 | 500 | 1.38 |
| V | 83-90 | 50 | 5000 | 13.8[b] |

[a] The reactors were inoculated with the fresh biomass at this period.

[b] Lactate concentration was increased by 10 times from day 85 onwards.

8.2.5. DGGE analysis

DGGE analysis has been carried out as described in a previous study (Jain et al., 2014a). The samples for DGGE analysis were taken on day 0 (inoculum), day 43 (end of period II) and day 90 (end of the study).

8.2.6. Analytics

The samples from the influent, effluent and the sampling ports were analyzed for the residual total Se concentration, after acidification, using an atomic absorption spectroscopy - Graphite Furnace (AAS-GF) (ThermoElemental Solaar MQZe GF95) and a Se lamp at 196.0 nm. The samples were acidified prior to measurement by adding a few drops of concentrated HNO_3 to see the concentration of total selenium. 15 ml of the original samples were centrifuged for 10 minutes at 37,000g, to separate the liquid phase from the pellets. The obtained pellets were re-solubilized in 15 ml of Milli-Q (18 MΩ*cm) water and acidified with 14.4 M HNO_3. Both the supernatant and re-solubilized samples were analyzed by AAS-GF to obtain the dissolved and elemental selenium concentration, respectively, in the effluent. Prior to AAS-GF analysis, the supernatant was acidified by adding a few drops of concentrated HNO_3. Selenate, nitrate and lactate in the effluent were determined by Ion Chromatography (IC) (Dionex ICS 1000), equipped with an AS4A column. The retention time of lactate, nitrate and selenate was 1.3, 3.9 and 11.3 min, respectively. Gas phase samples taken from the HNO_3 and NaOH traps were diluted two times with 6M NaOH (G1) and 14.4M HNO_3 (G2), respectively, to adjust the pH before AAS-GF analysis.

8.3. Results

8.3.1. Selenate reduction at 10 µM feed concentration (Period I)

Both UASB reactors were able to remove 10 µM selenate (790 µg Se L^{-1}) corresponding to a loading rate of 2.37 mg Se L^{-1} d^{-1}. After 25 days of operation, concentrations of total and dissolved selenium in the effluent were lower than the US EPA limit of 50 µg Se L^{-1} (USEPA, 2001), as shown by Figure 8.2. However, the performance of both reactors was not similar in the first few days of operation (0 - 9 days). The reduction of selenate was almost instantaneous in the mesophilic reactor, achieving a total and dissolved selenium removal efficiency of > 85% already after the first day of operation. The thermophilic reactor was able to achieve comparable efficiencies only after ~ 15 days. However, at the end of period I, the removal efficiency of both total and dissolved selenium was nearly identical (> 94%), when operating under mesophilic and thermophilic conditions.

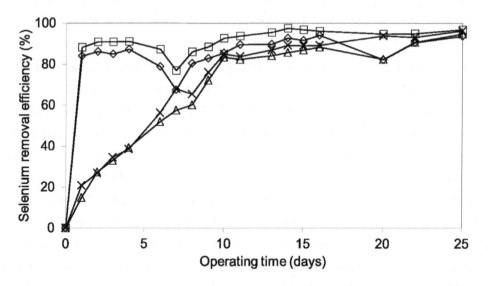

Figure 8.2. Comparison between the total and dissolved selenium removal efficiency in period I obtained under mesophilic (30°C, ◇ total, □ dissolved) and thermophilic (55°C, Δ total, × dissolved) conditions (influent concentration = 10 µM).

8.3.2. Selenate reduction at 50 µM fed concentration (Period II)

In period II, the fresh biomass was added to avoid interference of the trapped elemental selenium from period I in determining the real concentration of elemental selenium along the length of the reactor. In this period, a higher feed of selenate loading rates (5 times more than period I) did not affect the selenium removal efficiency of the mesophlic reactor and a high total selenium removal was achieved immediately. For the thermophilic reactor, removal of the total selenium started instantaneously, in contrast to when the influent selenate concentration was 5 times lower. During period II, the average removal efficiency of the total selenium was 85.0% under mesophilic conditions and 93.0% under thermophilic conditions. The dissolved selenium removal efficiency was 94.9% and 95.2% under mesophilic and thermophilic conditions, respectively (Figure 8.3a).

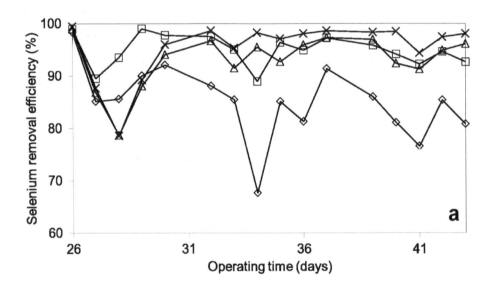

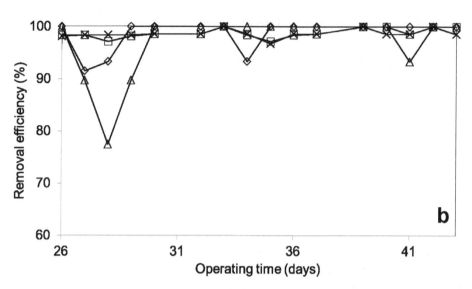

Figure 8.3. (a) Comparison between the total and dissolved selenium removal efficiency in period II obtained under mesophilic (30°C, ◇ total, □ dissolved) and thermophilic (55°C, Δ total, × dissolved) conditions (influent concentration = 50 μM). (b) Comparison between the selenate and lactate removal efficiency in period II obtained under mesophilic (30°C, ◇ selenate, □ lactate) and thermophilic (55°C, Δ selenate, × lactate) conditions (influent concentration = 50 μM).

For period II, after a few days of stabilization, selenate was not detected anymore in the influent (< 400 μg L^{-1}) until the end of the experiment. The removal efficiency was > 90% during the entire operation time in period II, excluding the start-up period (Figure 8.3b) for both the reactors. Lactate, used as carbon source and electron donor for the reduction of selenate with a concentration of 167 mg COD L^{-1}, was consumed by more than 99%.

8.3.3. Elemental selenium stratification along the length of the reactors

The elemental selenium concentration measured in each sampling port (S1, S2 and S3) of the mesophilic reactor was higher than the corresponding sampling port values measured in the thermophilic reactor (Figure 8.4a) during the entire period II of the reactor operations. The concentration of elemental selenium measured in the different ports of the thermophilic reactor did not change with the time, and it ranged between 50 and 140 µg L^{-1} in the different sampling ports. The concentrations of elemental selenium in the samples recovered from the mesophilic reactor were more unstable. It was higher on day 34, ranging between 800 to 1000 µg L^{-1} in the different sampling ports, which then started to decrease reaching a minimum (~400 µg L^{-1}) on day 39 and then increased and decreased again until the end of the experiment (day 43). The concentration of elemental selenium measured at sampling port S3 of the thermophilic reactor was every time higher than the concentrations measured at the other sampling ports.

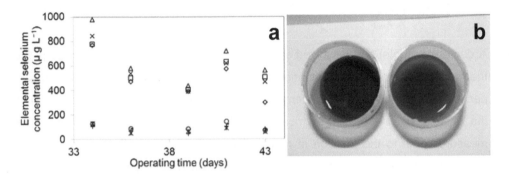

Figure 8.4. a) Concentration profiles of elemental selenium along the reactor height as a function of operational time. The concentration of elemental selenium measured in the samples recovered by sampling ports S1 (◇), S2 (□), S3 (Δ) and effluent (×) of the mesophilic reactor; concentration of elemental selenium measured in the samples recovered by sampling port S1 (*), S2 (○), S3 (+) and effluent (-) of thermophilic reactor. (b) Color of elemental selenium produced under mesophilic (left) and thermophilic (right) conditions.

The color of elemental selenium produced under mesophilic and thermophilic conditions were different from day 1 (Figure 8.4b). This was observed in all the ports as well as in the effluent. Red colored elemental selenium was observed in the mesophilic reactor, while grey colored elemental selenium was observed in the thermophilic reactor.

8.3.4. Effects of nitrate on selenate removal (Period III - V)

During period III (44 - 60 days), an addition of 6.2 mg L^{-1} (100 µM) of nitrate (NO_3^- / SeO_4^{2-} = 2) in the influent did not affect or rather slightly improve the selenium removal efficiency in the mesophilic reactor. Both the total (88 ± 3%) and dissolved (95 ± 2%) selenium removal efficiencies in the effluent of the mesophilic reactor was comparable to the concentrations measured prior to the addition of nitrate (Figure 8.5a, b). Selenate was never detected in the effluent of the mesophilic reactor for the whole period III (Figure 8.5c).

A decrease of total and dissolved selenium removal efficiency was observed in the thermophilic reactor after the addition of 100 µM of nitrate. During the first 4 days of period III (days 44 - 47), the concentration of both the total and dissolved selenium in the effluent steadily increased, rising to 8-9 times higher than the concentration measured prior to the addition of nitrate. The lower removal efficiency of ~68 and ~70% of total and dissolved selenium, respectively, was detected on day 56. The selenate concentration in the effluent of the thermophilic reactor followed the same trend of dissolved selenium. Nitrate was not detected in the effluent of the thermophilic reactor, suggesting that the concentration was lower than the detection limit of 1 mg L^{-1} (~ 16 µM). Surprisingly, nitrate was increasingly found in the effluent of mesophilic reactor, reaching a concentration of 30 µM (1.86 mg L^{-1}) on day 60 (Figure 8.5d). Lactate was completely consumed by both the reactors at 1.38 µM of feed concentration (Figure 8.5e).

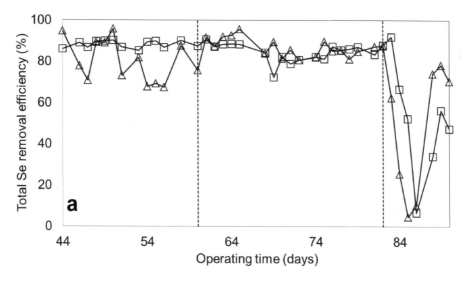

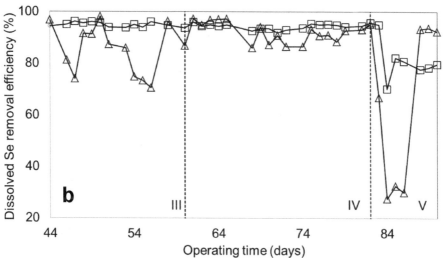

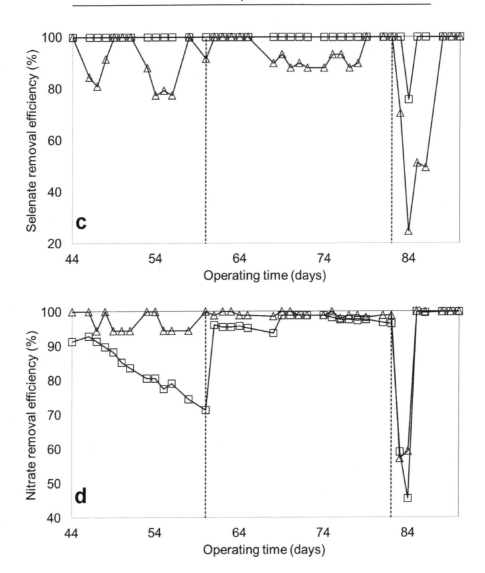

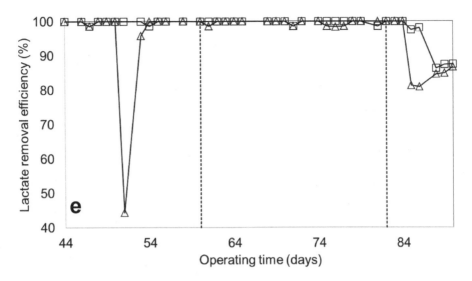

Figure 8.5. Removal efficiency (%) of total selenium (a), dissolved selenium (b), selenate (c), nitrate (d) and lactate (e) during periods III, IV and V (□ mesophilic, △ thermophilic).

During period IV (61-82), when the concentration of nitrate was increased to 31 mg L^{-1} (500 µM) (NO_3^- / SeO_4^{2-} = 10), the removal efficiency of total selenium under mesophilic conditions was lower than 4% as compared to the efficiency observed in period III. The removal efficiency of dissolved selenium remained unaffected, ranging between 92 and 96%. Selenate was never found in the effluent of the mesophilic reactor during period IV.

The thermophilic reactor was affected by the presence of 500 µM of nitrate when compared to the removal efficiencies observed in period II. However, when the total selenium removal efficiency was compared to those observed in period III, there was an increase of 7% in period IV with an average removal efficiency of 86.9%. The average dissolved selenium removal efficiency also increased by 7% when compared to period III, from 85.3 to 92.2% (Figure 8.5b). As observed previously in period III, nitrate was not detected in the effluent of the thermophilic reactor, while low nitrate concentrations < 2.5 mg L^{-1} (40 µM) were detected in the effluent of mesophilic reactor (Figure 8.5d). Selenate was detected at an average concentration of 217 µg L^{-1} in the thermophilic reactor (Figure 8.5c). Lactate was completely consumed under both the operating conditions (Figure 8.5e).

The concentration of nitrate in the influent was increased to 310 mg L^{-1} (5000 µM, NO_3^- / SeO_4^{2-} = 100) in period V. The removal efficiency of both total and dissolved selenium in the effluent of both the reactors decreased during days 83 – 85 as lactate was limiting. After a 10 times increase in the influent lactate concentration from 1.38 to 13.8 mM, the removal efficiency of both total and dissolved selenium increased until the end of the experiment. The removal efficiency of the total and dissolved selenium under mesophilic conditions on day 90 were ~ 48% (Figure 8.5a) and ~ 80% (Figure 8.5b), with a concentration of total and dissolved selenium equal to 2070 and 800 µg L^{-1}, respectively. On the same day, the removal efficiency of the total and dissolved selenium under thermophilic conditions were ~ 71% and ~ 92%, with a concentration of total and dissolved selenium equal to 1160 and 300 µg L^{-1}, respectively. It is interesting to note that nitrate was removed more than 99% in both the reactors in period V, after the increase in the lactate feed concentration (Figure 8.5d).

The removal efficiency of selenate under thermophilic and mesophilic decreased when lactate was limiting (days 83-85). However, selenate was not detected after an increase in lactate concentration under mesophilic operating conditions, but was detected at ~ 2000 µg L^{-1} for two more days (days 85-87) under the thermophilic conditions. After the increase in the lactate concentration from 1.38 mM to 13.8 mM, nitrate was almost completely removed (> 99%) under both operating conditions (Figure 8.5d). The removal efficiency of lactate, after a 10 times increase in the influent concentration, was > 80% under both operating conditions. The average removal efficiencies of total and dissolved selenium, selenate, nitrate and lactate in the different operating periods for mesophilic and thermophilic reactors are summarized in Table 8.6.

Table 8.6. Average removal efficiencies of total selenium, dissolved selenium, selenate, nitrate and lactate during different operating periods under mesophilic and thermophilic conditions. Influent concentrations: 10 µM L^{-1} (period I) and 50 µM L^{-1} (periods II, III, IV, V) of selenate; 100 µM L^{-1} (period III), 500 µM L^{-1} (period IV) and

5000 µM L^{-1} (period V) of nitrate; 1.389 mM L^{-1} (periods I, II, III and IV) and 13.89 mM L^{-1} (period V) of lactate

Operating condition	Period	Operating days	Total Se removal (%)	Diss. Se removal (%)	SeO$_4^{2-}$ removal (%)	NO$_3^-$ removal (%)	Lactate removal (%)
Mesophilic	I	1 - 25	85.9	91.7	-	-	-
	II	26 - 43	85.0	94.9	98.6	-	98.8
	III	44 - 60	88.3	95.1	99.9	83.0	99.8
	IV	61 - 82	84.6	94.3	99.9	97.0	99.8
	V	85 - 88[a]	39.5	79.8	99.9	99.9	91.4
Thermophilic	I	1 - 25	66.7	70.1	-	-	-
	II	26 - 43	93.0	95.2	96.9	-	98.7
	III	44 - 60	80.1	85.3	90.0	97.0	95.3
	IV	61 - 82	86.9	92.2	94.5	99.1	99.6
	V	85 - 88[a]	47.6	68.4	80.1	99.9	83.8

Note: [a]Days 83 and 84 were removed from the table because lactate was limiting under those conditions

8.3.5. Speciation of selenium in the effluent

Figure 8.6 shows the average selenium speciation in the effluent of both the reactors at all the periods. In the mesophilic reactor, the elemental selenium fraction was 60% in average and thus higher than the other fractions (dissolved selenium and selenate). In contrast, the average elemental selenium fraction during the various operational periods in the thermophilic reactor was only 30%. Most of the dissolved selenium was found as selenate in the effluent of the mesophilic reactor, while the percentage of other unknown dissolved selenium species were higher in the effluent of the thermophilic reactor.

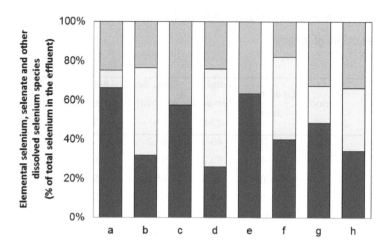

Figure 8.6. Percentage of total selenium found as elemental selenium (■), selenate (□) and other dissolved selenium compounds (▦) during period II (a), III (c), IV (e) and V (g) in the effluent of mesophilic reactor and during period II (b), III (d), IV (f) and V (h) in the effluent of thermopilic reactor.

8.3.6. DGGE analysis of anaerobic granules at the end of periods II and V

A DGGE analysis was carried out to observe the rise of the different microbial populations due to the exposure of selenium, nitrate as well as the effect of temperatures (Figure 8.7a). The differences in the bands in the DGGE analysis (marked by square) suggest the change in the microbial population in the anaerobic granules fed with selenate at the end of period II. At the end of period V, there are differences in the bands suggesting the further change in the microbial community of both reactors upon the addition of nitrate in the feed of the reactor (Figure 8.7a). DGGE analysis of the methanogens revealed the presence of methanogens in the inoculum sludge. These methanogens survived the period II and were present in the mesophilic reactor (Figure 8.7b). However, methanogens were completely absent in the thermophilic reactor as well upon addition of nitrate in the fed in both the mesophilic and thermophilic reactor (Figure 8.7b).

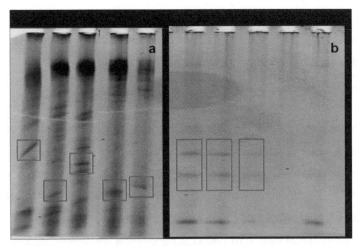

Figure 8.7. DGGE analysis of (a) bacteria and (b) methanogens (*Archae*) for the inoculum anaerobic granules (L1), mesophilic anaerobic granules at the end of period II (L2) and V(L4) and thermophilic anaerobic granules at the end of period II (L3) and V (L5).

8.4. Discussion

8.4.1. Biological removal of selenium in UASB reactors

This study demonstrated that operating UASB reactor at thermophilic conditions are more suited for the better removal of the total selenium when the competing electron acceptor nitrate is absent (Period II, Figure 8.3). Both reactors achieved a similar total dissolved selenium removal efficiency in period II, which suggests that selenate reduction to elemental selenium can be carried out in an UASB reactor under thermophilic conditions as well (Lenz et al., 2008a). The higher total selenium removal efficiency in the thermophilic reactor was due to the higher elemental selenium nanoparticles concentration in the effluent of mesophilic reactors (6% of total fed selenium) as compared to the thermophilic reactor (3% of total fed selenium). The absence of selenium in the gas traps for both the reactors suggests that the major selenate removal mechanism in both reactors was selenate reduction to elemental selenium and its subsequent retention in the bioreactor. The elemental selenium formed in the UASB reactor could be formed intracellularly or trapped within the anaerobic granules in their elemental form or as metal selenide (Lenz et al., 2008a). Another possibility is that the low upflow velocity (27.3 cm h^{-1}) allows selenium particles to settled down following interactions with wastewater cations that decrease the colloidal stability of the Se nanoparticles (Buchs et al., 2013; Jain et al., 2015; Lenz et al., 2008a; Pi et al., 2013).

It is also interesting to point out that the size and shape of the elemental selenium precipitates is very different in the mesophilic and thermophilic reactors (Figure 8.4b). Elemental selenium formed under thermophilic conditions, as used in this study, leads to the formation of trigonal selenium nanowires (BioSeNWs) with a median diameter of 40-50 nm and length of micrometer (Jain et al., 2014),. In contrast, the elemental selenium produced under the mesophilic conditions is in the form of spherical particles with a median diameter of 180nm (BioSeNPs) (Jain et al., 2015). Both the BioSeNPs and BioSeNWs display nevertheless similar ζ-potential profiles (Jain et al., 2014; 2014b), the difference in their shape would lead to different interactions with cations leading to different settling properties.

Selenate reduction under anaerobic conditions normally follows the pathway leading to selenite and then to elemental selenium. This reduction can take place either intracellularly or extracellularly depending on the type of microorganisms (Hockin and Gadd, 2006; Kessi and Hanselmann, 2004; Li et al., 2014a; Tomei et al., 1995). Thus, the presence of elemental selenium in the effluent would be strongly influenced by the evolution of the microbial community. Indeed, the DGGE analysis showed large differences in the microbial community structure of anaerobic granules under mesophilic and thermophilic conditions (Figure 8.7).

It is interesting to note that the at 10 µM selenate feed, 9 days longer adaptation time was required for the thermophilic reactor to achieve selenium removal efficiencies that are comparable to those obtained under mesophilic conditions. Surprisingly, this was not observed when the same inoculum sludge was used, but when using a feed selenate concentration of 50 µM. This can be attributed to the fact that the selenate reduction rate in a continuous reactor has a strongly dependence by the selenium loading (Takada et al., 2008), making the reduction of selenate faster. At the lower selenate feed concentrations investigated, the longer adaptation time can be attributed to the time required for acclimatization of the mesophilic inoculum to the thermophilic conditions (Khemkhao et al., 2012; Li et al., 2014).

8.4.2. Effect of nitrate on selenium removal

Lai et al. (2014) demonstrated that the reduction of selenate is dramatically inhibited by the presence of nitrate at a surface loading higher than 1.14 g N m^{-2} d^{-1} (10 mg L^{-1} fed in the reactor) and that the selenate-reducing microbial community can be reshaped by the presence of nitrate in a hydrogen based membrane biofilm reactor (MBfR). On the contrary, in this study, the selenate reduction was not affected by the presence of nitrate until a nitrate to selenate ratio of 100 (31 mg L^{-1} fed in the reactor) under mesophilic operating condition (Table 8.6). This might be due to the fact that the UASB reactors operated in this study were not under electron donor limitation. The other possible reasons is the rise of specialist selenate reducing bacteria which has also been indicated by the DGGE analysis (Figure 8.7) and also observed by Lenz et al. (2009). The presence of residual nitrate in the effluent of the mesophilic reactor (Figure 8.5d) suggests the possibility of the rise of specialist selenate reducing bacteria which were not affected by the presence of nitrate.

The presence of nitrate affected the selenate removal under thermophilic conditions, which was evident from the detection of selenate in the effluent and a near complete removal of nitrate from the effluent. This further suggests that the reason for selenate removal in presence of nitrate under mesophilic conditions might be due to development of specialist selenate reducing bacteria and not due to unlimited electron donor. The DGGE analysis showed the different microbial population at the end of period II for both the mesophilic and thermophilic reactors (8.7).

When the feed concentration of nitrate was increased from 100 to 500 µM L^{-1}, there was an increase in the total selenium concentration in the effluent of the mesophilic reactor (period IV). However, the dissolved selenium concentration almost remained the same (Table 8.6). This suggests the excess release of elemental selenium in the effluent of the mesophilic reactor in period IV. This can be due to the change in the metabolic pathways of selenate reduction in the microorganism leading to extracellular production of elemental selenium. Indeed, growth of the microorganisms in the nitrate followed by selenite reduction resulted in extracellular production of elemental selenium in *Sulfurospirillum barnesii*, *Bacillus selenitireducens*, and *Selenihalanaerobacter shriftii* (Oremland et al., 2004). The increase in the difference between total and dissolved selenium was not observed in periods III and IV for the thermophilic reactor. This suggests that thermophilic reactors are less susceptible to release of elemental selenium when nitrate is fed. Though the selenium removal efficiency is affected when nitrate is fed in the thermophilic reactor, the increase in total and dissolved selenium removal efficiency in period IV (from period III) to 92.2 and 86.9%, respectively, suggests the possibility of acclimatizing the thermophilic reactor to achieve a desirable total and dissolved selenium removal efficiency even in the presence of nitrate.

8.4.3. Selenium speciation in the effluent

The concentration of elemental form of selenium was consistently less in the effluent of the thermophilic reactor than in the mesophilic reactor (Figure 8.6). This was further corroborated with the higher total selenium concentration in the mesophilic reactor as compared to thermophilic reactor in period II of reactor operations. The presence of volatile selenium compounds in the gas traps was negligible at 30°C, as observed in earlier studies (Lenz et al., 2008a; Lenz al., 2008b). Surprisingly, it was also negligible at 55°C, suggesting that the vaporization of alkylated compounds is not related to the operational conditions. Dimethylselenide and dimethyldiselenide can exist also as dissolved alkylated compounds (Lenz et al., 2008b), contributing to the concentration of dissolved selenium that leaves the UASB reactor with the effluent. Indeed, the dissolved selenium concentration was greater than the selenate concentration in both reactors, suggesting the presence of dimethylselenide and dimethyldiselenide. Also, no selenide was observed in the reactor operation as the concentration of selenium in the gas trap 2 was negligible. However, it is plausible that selenide was formed and then quickly oxidized to elemental selenium or formed metal selenide and thus, would be virtually undetectable.

8.5. Conclusions

This study demonstrated that at 50 µM (3950 µg L^{-1}) feed of selenate, average total selenium removal efficiency was higher (+ 8%) under thermophilic conditions, because of the higher concentration of elemental selenium in the effluent of the mesophilic reactor which were also confirmed by the stratification results (Figure 8.3a). Elemental selenium produced under mesophilic and thermophilic conditions was different in color most likely due to a different crystalline structure. The addition of 100 and 500 µM L^{-1} of nitrate affected the reduction of selenate only under thermophilic conditions, highlighting the development of different microorganisms under mesophilic and thermophilic conditions, confirmed by DGGE analysis. However, the total and dissolved selenium removal efficiency improved by 6% when the nitrate feed was increased from 100 to 500 µM L^{-1} in the thermophilic reactor suggesting the possibility of development of specialist selenate reducers. Interestingly, the increase in the concentration of nitrate feed led to a larger release of elemental selenium in the effluent under mesophilic conditions. This phenomenon was not observed under thermophilic conditions.

8.6. References

Buchs, B., Evangelou, M.W.H., Winkel, L.H.E., Lenz, M. 2013. Colloidal Properties of Nanoparticular Biogenic Selenium Govern Environmental Fate and Bioremediation Effectiveness. Environmental Science & Technology, 47(5), 2401-2407.

Cantafio, A.W., Hagen, K.D., Lewis, G.E., Bledsoe, T.L., Nunan, K.M., Macy, J.M. 1996. Pilot-Scale Selenium Bioremediation of San Joaquin Drainage Water with Thauera selenatis. Appl Environ Microbiol, 62(9), 3298-303.

Frankenberger, W.T., Jr., Amrhein, C., Fan, T.W.M., Flaschi, D., Glater, J., Kartinen, E., Jr., Kovac, K., Lee, E., Ohlendorf, H.M., Owens, L., Terry, N., Toto, A. 2004. Advanced Treatment Technologies in the Remediation of Seleniferous Drainage Waters and Sediments. Irrigation and Drainage Systems, 18(1), 19-42.

Hockin, S., Gadd, G.M. 2006. Removal of selenate from sulfate-containing media by sulfate-reducing bacterial biofilms. Environmental Microbiology, 8(5), 816-826.

Jain, R., Jordan, N., Schild, D., van Hullebusch, E.D., Weiss, S., Franzen, C., Farges, F., Hübner, R., Lens, P.N.L. 2015. Adsorption of zinc by biogenic elemental selenium nanoparticles. Chemical Engineering Journal, 260(0), 855-863.

Jain, R., Jordan, N., Kacker, R., Weiss, S., Hübner, R., Kramer, H., van Hullebusch, E.D., Farges, F., Lens, P.N.L, 2014. Biogenic synthesis of elemental selenium nanowires *(in preparation)*.

Jain, R., Matassa, S., Singh, S., Hullebusch, E.D. Van, Esposito, G., Farges, F., Lens, P.N.L., 2014a. Reduction of selenite to elemental selenium nanoparticles by activated sludge under aerobic conditions. Biochem. Eng. J. *(submitted)*.

Jain, R., Jordan, N., Weiss, S., Foerstendorf, H., Heim, K., Kacker, R., Hübner, R., Kramer, H., Hullebusch, E.D. Van, Farges, F., Lens, P.N.L., 2014b. Extracellular polymeric substances (EPS) govern the surface charge of biogenic elemental selenium nanoparticles (BioSeNPs). Environ. Sci. Tech. *(Submitted)*.

Kessi, J., Hanselmann, K.W. 2004. Similarities between the Abiotic Reduction of Selenite with Glutathione and the Dissimilatory Reaction Mediated by Rhodospirillum rubrum and Escherichia coli. Journal of Biological Chemistry, 279(49), 50662-50669.

Khemkhao, M., Nuntakumjorn, B., Techkarnjanaruk, S., Phalakornkule, C. 2012. UASB performance and microbial adaptation during a transition from mesophilic to thermophilic treatment of palm oil mill effluent. Journal of Environmental Management, 103(0), 74-82.

Lai, C.Y., Yang, X., Tang, Y., Rittmann, B.E., Zhao, H.P. 2014. Nitrate shaped the selenate-reducing microbial community in a hydrogen-based biofilm reactor. Environ Sci Technol, 48(6), 3395-402.

Lee, J.H., Han, J., Choi, H., Hur, H.G. 2007. Effects of temperature and dissolved oxygen on Se(IV) removal and Se(0) precipitation by Shewanella sp. HN-41. Chemosphere, 68(10), 1898-905.

Lenz, M., Enright, A.M., O'Flaherty, V., van Aelst, A.C., Lens, P.N. 2009. Bioaugmentation of UASB reactors with immobilized Sulfurospirillum barnesii for simultaneous selenate and nitrate removal. Appl Microbiol Biotechnol, 83(2), 377-88.

Lenz, M., Hullebusch, E.D.V., Hommes, G., Corvini, P.F., Lens, P.N. 2008a. Selenate removal in methanogenic and sulfate-reducing upflow anaerobic sludge bed reactors. Water research, 42(8), 2184-2194.

Lenz, M., Lens, P.N.L. 2009. The essential toxin: The changing perception of selenium in environmental sciences. Science of The Total Environment, 407(12), 3620-3633.

Lenz, M., Smit, M., Binder, P., van Aelst, A.C., Lens, P.N. 2008b. Biological alkylation and colloid formation of selenium in methanogenic UASB reactors. J Environ Qual, 37(5), 1691-700.

Li, D.B., Cheng, Y.Y., Wu, C., Li, W.W., Li, N., Yang, Z.C., Tong, Z.H., Yu, H.Q. 2014a. Selenite reduction by Shewanella oneidensis MR-1 is mediated by fumarate reductase in periplasm. Sci Rep, 4, 3735.

Li, X.-k., Ma, K.-l., Meng, L.-w., Zhang, J., Wang, K. 2014b. Performance and microbial community profiles in an anaerobic reactor treating with simulated PTA wastewater: From mesophilic to thermophilic temperature. Water Research, 61(0), 57-66.

NSMP. 2007. Identification and assessment of selenium and nitrogen treatment technologies and best management practices. Available at http://www.ocnsmp.com/library.asp

Oremland, R.S., Herbel, M.J., Blum, J.S., Langley, S., Beveridge, T.J., Ajayan, P.M., Sutto, T., Ellis, A.V., Curran, S. 2004. Structural and spectral features of selenium nanospheres produced by Se-respiring bacteria. Appl Environ Microbiol, 70(1), 52-60.

Pi, J., Yang, F., Jin, H., Huang, X., Liu, R., Yang, P., Cai, J. 2013. Selenium nanoparticles induced membrane bio-mechanical property changes in MCF-7 cells by disturbing membrane molecules and F-actin. Bioorganic & Medicinal Chemistry Letters, 23(23), 6296-6303.

Roest, K., Heilig, H.G., Smidt, H., de Vos, W.M., Stams, A.J., Akkermans, A.D. 2005. Community analysis of a full-scale anaerobic bioreactor treating paper mill wastewater. Syst Appl Microbiol, 28(2), 175-85.

Soda, S., Kashiwa, M., Kagami, T., Kuroda, M., Yamashita, M., Ike, M. 2011. Laboratory-scale bioreactors for soluble selenium removal from selenium refinery wastewater using anaerobic sludge. Desalination, 279(1–3), 433-438.

Takada, T., Hirata, M., Kokubu, S., Toorisaka, E., Ozaki, M., Hano, T. 2008. Kinetic study on biological reduction of selenium compounds. Process Biochemistry, 43(11), 1304-1307.

Tam, K., Ho, C.T., Lee, J.H., Lai, M., Chang, C.H., Rheem, Y., Chen, W., Hur, H.G., Myung, N.V. 2010. Growth mechanism of amorphous selenium nanoparticles synthesized by Shewanella sp. HN-41. Biosci Biotechnol Biochem, 74(4), 696-700.

Tomei, F., Barton, L., Lemanski, C., Zocco, T., Fink, N., Sillerud, L. 1995. Transformation of selenate and selenite to elemental selenium byDesulfovibrio desulfuricans. Journal of Industrial Microbiology, 14(3-4), 329-336.

USEPA. 2001. Final report selenium treatment / removal alternatives demonstration project - Available at
http://www.epa.gov/ORD/NRMRL/std/mtb/mwt/a3/mwtp191.pdf.

Zhang, Y., Wang, J., Amrhein, C., Frankenberger, W.T. 2005. Removal of Selenate from Water by Zerovalent Iron. J. Environ. Qual., 34(2), 487-495.

CHAPTER 9

Reduction of selenite to elemental selenium nanoparticles by activated sludge under aerobic conditions

This chapter will be submitted as

Jain, R., Matassa, S., Singh, S., Hullebusch, E.D. Van, Esposito, G., Lens, P.N.L., 2015. Reduction of selenite to elemental selenium nanoparticles by activated sludge under aerobic conditions.Process Biochem (*to be submitted*)

Abstract:

This study proposes a one-step process for total selenium removal using activated sludge, where selenite is reduced, while the produced colloidal elemental selenium nanoparticles (BioSeNPs) are entrapped in the activated sludge flocs. Glucose as carbon source gave 2.9 and 6.8 times higher total selenium removal at neutral pH and 30 oC as compared to lactate and acetate at 2.0 g L-1 COD. Total selenium removal efficiencies of 79±3 and 86±1% were achieved, respectively, in shake flasks and batch reactors at a dissolved oxygen (DO) > 4.0 mg L-1 and 30 oC when fed with 172 mg L−1 Na2SeO3 and 2.0 g L−1 chemical oxygen demand (COD) of glucose. Continuous reactors operated at neutral pH and 30 oC removed 33.98 and 36.65 mg of total selenium per g of total suspended solids (TSS) at TSS concentrations of 1300 and 3000 mg L−1, respectively. However, the operated reactors crashed upon continuous feeding of selenium after 10-20 days at the applied loading rates, most likely due to toxicity of selenite to the aerobic bacteria.

Keywords: aerobic processes, selenium, nanoparticles, waste-water treatment, bioreactors, dissolved oxygen, one-step process

Graphical abstract:

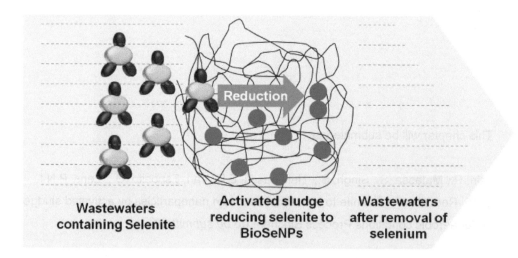

Wastewaters containing Selenite

Activated sludge reducing selenite to BioSeNPs

Wastewaters after removal of selenium

Reduction

9.1. Introduction

The toxicity of selenate (SeO_4^{2-}) and selenite (SeO_3^{2-}) to the environment, including humans, aquatic life and animals, has been well documented in the past few decades (Lenz and Lens, 2009). Anaerobic biological reduction converts selenate (SeO_4^{2-}) and selenite (SeO_3^{2-}) to the more stable elemental selenium at a reasonable cost and an acceptable remediation efficiency (Cantafio et al., 1996; Lenz et al., 2008). However, the elemental selenium formed after microbial reduction is in the form of dispersed colloidal elemental selenium nanoparticles (BioSeNPs). These BioSeNPs are present in the effluent of selenate and selenite reducing upflow anaerobic sludge blanket (UASB) reactors and can re-oxidize in the environment (Buchs et al., 2013). These BioSeNPs have to be removed prior to discharge, thus an additional coagulation step is required leading to increased treatment costs (Buchs et al., 2013; Staicu et al., 2014).

Selenite reduction through the dissimilatory pathway under anaerobic conditions, as present in an UASB reactor, results in the extracellular production of BioSeNPs resulting in higher total selenium effluent concentrations (Li et al., 2014). Selenium tolerant bacteria can reduce selenite (SeO_3^{2-}) to BioSeNPs through detoxification mechanisms, in which BioSeNPs remain entrapped within the biomass under aerobic and microaerobic condition (Dhanjal and Cameotra, 2010; Tejo Prakash et al., 2009). This provides an alternative pathway of total selenium removal, without the need of a second step to remove the BioSeNPs as required in the anaerobic removal.

The reduction of selenium oxyanions under aerobic conditions has been documented for axenic cultures (Jain et al., 2014). In contrast to axenic cultures, activated sludge may have better trapping ability of BioSeNPs due to its flocs structure. The entrapment of selenium in the biomass is attractive as the BioSeNPs containing activated sludge can be used to recover the selenium, can be applied as a slow release selenium fertilizer (Haug et al., 2007), or used for treating mercury contaminated soils (Johnson et al., 2008).

This study aimed to investigate the removal of the total dissolved selenium by reduction of selenite to elemental selenium and entrapping of the formed BioSeNPs in the activated sludge flocs. Shake flask experiments were conducted to identify the optimum carbon source, dosage and temperature for total selenium removal from the aqueous phase. The effect of dissolved oxygen (DO levels) concentration on the total selenium removal from the aqueous phase was studied in batch reactors. A continuously operated activated sludge reactor with sludge recycle was operated to evaluate the total selenium removal by a continuously fed system. Population dynamics in the activated sludge biomass were characterized by denaturing gradient gel electrophoresis (DGGE) analysis.

9.2. Materials and methods

9.2.1. Source of biomass and growth medium

Activated sludge reducing selenite under aerobic conditions was collected from a full scale domestic wastewater treatment plant in Harnaschpolder, The Netherlands (Karya et al., 2013). A synthetic mineral medium for shake flask experiments, batch reactors and continuous reactor was prepared by dissolving 600 mg NH_4Cl, 200 mg $MgCl_2 \cdot 6H_2O$, 20 mg $CaCl_2 \cdot 2H_2O$ and 1 mL micronutrient solution in 1.0 L of Milli-Q (18 $M\Omega^*cm$) water. For the continuous reactor, 23 mg Na_2HPO_4 was also added in 1.0 L of Milli-Q water.

9.2.2. Batch flask experiments

Shake flask experiments to determine the optimum carbon source, dosage and temperature for total selenium removal were carried out. The tests were done in 300 mL Erlenmeyer flasks, with a working volume of 100 mL. Each flask was inoculated with activated sludge for a resulting total suspended solids (TSS) of 500±50 mg L^{-1}. Mixing in the shake flasks were carried out using rotary shakers at 150 rpm. The initial pH was manually adjusted to values around 7.5, and if needed sodium carbonate (Na_2CO_3) was used to buffer the medium.

178

Selenite, in the form of sodium selenite (Na_2SeO_3), was added until achieving a final selenium concentration of 79 mg L^{-1} (1mM). Glucose (glucose monohydrate - $C_6H_{12}O_6 \cdot H_2O$), lactate (sodium lactate - $NaC_3H_5O_3$) and acetate (sodium acetate - $NaC_2H_4O_2$) were tested with different chemical oxygen demand (COD) concentrations of 0.5, 1.0 and 2.0 g L^{-1}. The experiments with glucose as the carbon source were incubated at 15±1, 30±1 and 45±1 °C, while experiments with lactate and acetate as the carbon source were incubated at 30±1 °C. For the calculations of the amount of total selenium removed per unit of biomass, the TSS concentration measured at the end of each experiment was used. All the shake flasks experiments were done in triplicate.

9.2.3. Batch reactors

Batch reactors were used to close the mass balance of the fed selenium at two different DO levels. The reactors consisted of 1 L glass vessel with a working volume of 500 mL. Each reactor was closed and connected to two gas traps filled with 65% concentrated nitric acid (100 mL and 50 mL respectively) in order to trap the volatilized selenium species (dimethyl selenide and dimethyl diselenide) potentially generated inside the reactors as described previous (Winkel et al., 2010).

All reactors were inoculated with the activated sludge (for a resulting TSS of 1300±100 mg L^{-1}) in the same mineral medium as in the batch flasks, and supplemented with glucose (2.0 g L^{-1} COD) and sodium selenite (172 mg L^{-1} or 1mM). The pH was controlled by addition of sodium carbonate, as well as by manual adjustment to values around 7.5 along the experiment. DO levels were varied by controlling the airflow rate through a sparger and DO levels were measured every 4 hours. The lower air flow rate of 2-5 L h^{-1} and higher air flow rate of 10-15 L h^{-1} were maintained. Batch reactors were operated in duplicates and if the difference in measurements were more than 10%, experiments were repeated. The average values are presented in the figures.

9.2.4. Continuous reactors

A continuous flow reactor with complete sludge recycle and two gas traps was used to assess the selenium removal efficiency of a continuous activated sludge system (Figure 9.1). A synthetic wastewater with glucose (1 g L^{-1}) and sodium selenite (17.2 mg L^{-1} or 0.1 mM) was pumped into a 1 L glass reactor operating at a hydraulic retention time (HRT) of 8 hours. The reactor was operated at 30±1 °C, mixed with a magnetic stirrer and aerated with air flow of 20-25 L h^{-1} through sparger to keep the DO levels > 4.0 mg L^{-1}. The reactor was operated at a TSS concentration of 1300±100 mg L^{-1} for the first 28 days (Periods I to III), followed by re-inoculation with biomass at a TSS concentration of 3000±100 mg L^{-1}. The reactor was operated at these conditions for another 38 days (Periods IV to VI) (Table 1). More details about the continuous reactor operational parameters are available in Table S1 of Appendix 4.

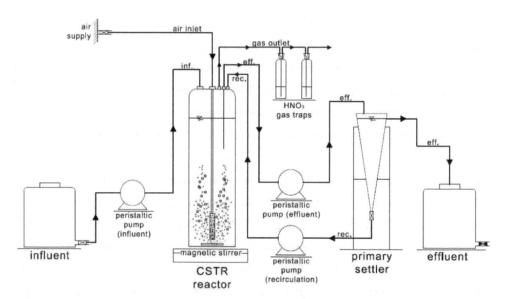

Figure 9.1. Configuration of the continuous activated sludge reactor with recirculation of the sludge and connected to two gas traps.

Table 9.1. Different operating conditions of the continuous activated sludge reactor with sludge recycle. Please note that the * represents the addition of fresh activated sludge at the start of period IV

Parameter	Experimental periods					
	I	II	III	IV*	V	VI
Days	0-10	11-21	22-29	30-36	37-56	57-66
Influent COD [mg L^{-1}]	1000	1000	1000	1000	1000	1000
Influent SeO$_3^{2-}$ [mg L^{-1}]	0	23.7	0	0	23.7	0
COD / Se ratio	-	126.6	-	-	126.6	-
TSS [mg L^{-1}]	4500	1300	1500	4300	3000	2900
TSS / Se ratio	-	54.9	-	-	126.6	-

9.2.5. Analytical methods

The total selenium concentration from the supernatant, activated sludge and gas phase were measured using Inductively Coupled Plasma Mass Spectrometry (ICP-MS) (Jain et al., 2015). BioSeNPs were separated from the aqueous phase by centrifugation (37,000 g, 15 minutes, Hermle Z36HK). The supernatant and pellet constitute the dissolved selenium and BioSeNPs, respectively, and their concentrations were determined by dissolving them in concentrated HNO$_3$ prior to selenium concentration determination using ICP-MS. Trapped selenium concentration in the activated sludge flocs was measured after addition of HNO$_3$, followed by microwave destruction of the biomass and then measuring the selenium concentration. Selenium concentrations in the gas traps were measured after appropriate dilutions. The COD and total suspended solids measurements were carried out using the standard methods (APHA, 2005). Transmission electron microscopy-Energy Disperse X-ray spectroscopy was carried on diluted samples as described in Cosmidis et al. (2013) (more details in Appendix 4). The DNA extraction for DGGE analysis of activated sludge fed with and without sodium selenite and incubated for 7 days in the batch reactors at neutral pH, 2.0 g L^{-1} COD (glucose) and DO > 4 mg L^{-1} was carried out using FAST DNA SPIN kit from MP Biomedicals, USA as reported in Ahammad et al. (Ahammad et al., 2013). The details of polymerase chain reaction (PCR) amplification of DNA including used primers are described in Appendix 4 (Table S2 in Appendix 4). DGGE was carried out as described in Muyzer et al. (Muyzer et al., 1993) (more details in Appendix 4).

9.3. Results

9.3.1. Selenite reduction in batch flask experiments

Figures 9.2a shows the developed red color when activated sludge is incubated with 172 mg L^{-1} (1 mM) sodium selenite at 30 °C in the shake flasks, suggesting the reduction of selenite to elemental selenium or BioSeNPs. TEM-EDXS confirm the presence of spherical particles (Figure 9.2b1) in the activated sludge flocs, comprising of selenium (Figure 9.2b2). The Cu signals observed are due to the use of Cu grid for holding the samples.

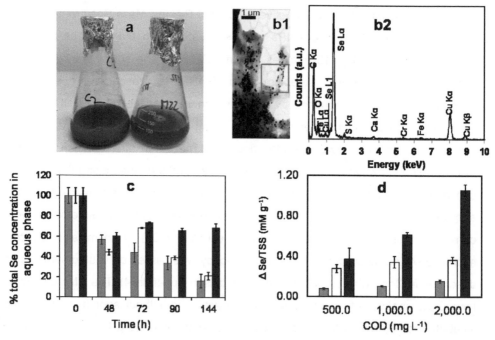

Figure 9.2. (a) Activated sludge trapping red elemental selenium (left) and control activated sludge fed no selenite (right); (b1) TEM image of the activated sludge trapping BioSeNPs and (b2) corresponding EDX spectrum of marked square of the sludge sample in b1; (c) % total selenium concentration as a function of time in the

supernatant of the activated sludge in shake flask experiments performed at 15 (■), 30 (□) and 45 °C (■) and 79 mg L^{-1} (1mM) of initial selenium concentrations and (d) Total selenium removal per g of TSS in the supernatant with increasing COD concentration of acetate (■), lactate (□) and glucose (■) in shake flask experiments.

The shake flask study at 2.0 g COD L^{-1} (glucose as carbon source) and different temperatures showed that the total selenium removal efficiency from aqueous phase was 84±7, 79±3 and 31±4% respectively, at 15, 30 and 45 °C, suggesting better total selenium removal at lower temperatures by the activated sludge (Figure 9.2c). No selenite reduction was observed in the abiotic control experiments with killed and without biomass (data not shown), confirming the selenite reduction is mediated by microbial activity.

Glucose removed total selenium 2.9 and 6.8 times higher from the liquid phase per g of TSS compared to lactate and acetate, respectively, at 2.0 g L^{-1} COD, suggesting that glucose was the best carbon source tested in this study (Figure 9.2d). The selenium removal efficiency from the liquid phase per g TSS at 2.0 g COD L^{-1} was, respectively, 2.8 and 1.7 times higher than at 0.5 and 1.0 g L^{-1} COD with glucose as the carbon source (Figure 9.2d). No detectable growth in the TSS was observed for all experimental conditions.

9.3.2. Selenite reduction in aerated batch reactors

The lower air flow rate (2-5 L h^{-1}) resulted in DO concentrations of 3.0 - 4.0 mg L^{-1} for entire duration, except for a few hours when the DO dropped to 0.1 mg L^{-1} (Figure 9.3a). The higher air flow rate (10-15 L h^{-1}) maintained DO levels consistently exceeding 6 mg L^{-1}, while the DO level was lowered to 4 mg L^{-1} only for a few hours In the aerated batch reactors, with DO levels always exceeding 4.0 mg L^{-1} (Figure 9.3a), the BioSeNPs (colloidal elemental selenium fraction) after 96 h of incubation at 30 °C constitutes only 3±1% of the total selenium added as compared to 17±1% in the reactor with DO levels below 4.0 mg L^{-1} (Figure 9.3b). The total selenium removal from the aqueous phase was 86±1% and 67±2% for reactors operating with DO levels, respectively, higher and lower than 4.0 mg L^{-1} (Figure 9.3b). The dissolved selenium removal efficiency for both DO levels was similar: 90±1% for higher DO levels compared to 85±3% for lower DO levels (Figure 9.3b).

This was further confirmed by the excess trapping of selenium in the biomass at higher DO levels: 65±1% at DO levels > 4.0 mg L^{-1} as compared to 48±2 % of total selenium at DO levels lower than 4.0 mg L^{-1} (Figure 9.3c). At both DO levels, COD removal profiles were similar (Figure 9.3b).

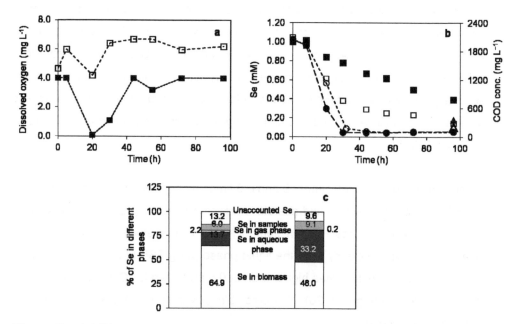

Figure 9.3. (a) Dissolved oxygen levels with time at high (□) and low (■) air flow rate, (b) Evolution of total selenium (□,■) and COD (—o—, -•-) concentration with time and dissolved selenium concentration at the end of incubation (△,▲) at DO levels always >4.0 mg L^{-1}: open symbols and at DO levels<4.0 mg L^{-1}: closed symbols, and (c) Mass balance: transfer of total fed selenium in the aqueous phase to biomass, gas traps, samples, remaining total selenium in the aqueous phase and unaccounted selenium.

9.3.3. Continuous operated activated sludge reactor

The pH and DO of the continuous reactor was near neutral (7-7.5) and always greater than 4 mg L^{-1} (Figure S1 in Appendix 4 for period I-III). In period I (Figure 9.4), no selenite was fed and a 90% COD removal efficiency was achieved. The TSS concentration increased to 2500 mg L^{-1} due to the biomass growth. On day 10

(Period II), the TSS concentration was set to 1300 mg L^{-1} in order to apply the same operative conditions as used in the batch reactors. In the period II, 77% removal of total selenium (23.7 mg L^{-1} d^{-1} or 0.3 mM d^{-1} selenium fed in form of selenite) was achieved in the first 2 days. The differences in the DGGE gel of the activated sludge fed with and without selenite in the batch reactors (Figure 9.5) suggest that a new population, most probably selenite reducing bacteria would have developed in the continuous reactors as well. However, both the total selenium and COD removal dropped to almost 0 and 20%, respectively, on the 21st day of operation, suggesting that selenite toxicity deteriorated the activated sludge performance and the newly developed selenium degrading species could not completely detoxify the selenite loading rate.

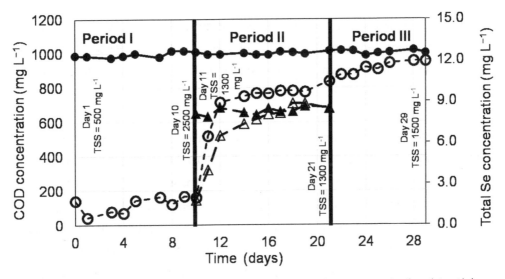

Figure 9.4. Evolution of the COD (•, ○) and total selenium concentration (▲, Δ) in the influent (closed symbols) and effluent (open symbols) in a continuously aerated activated sludge reactor with complete sludge recycle at a TSS of 1300 mg L^{-1}, pH 7.3 and DO >4.0 mg L^{-1}.

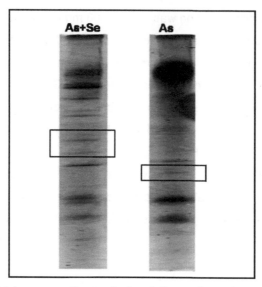

Figure 9.5. Microbial community analysis of the activated sludge fed by DGGE. Activated sludge fed with 172 mg L^{-1} of sodium selenite (AS+Se) and activated sludge (AS) without sodium selenite after 7 days of incubation with 2.0 g L^{-1} of glucose.

In period III, the selenite feed was stopped, however, the reactor could not recover. On the 28th day of reactor operations, the COD removal was still below 10%, suggesting an irreversible inactivation of the biomass. When the reactor was re-inoculated with fresh biomass at a higher TSS concentration (3000 mg L^{-1}; period IV-VI), a similar performance was observed (Figure S2a in Appendix 4). The neutral pH and DO levels greater than 3 mg L^{-1} were also maintained during the period V of the continuous reactor operations (Figure S2b in Appendix 4). The COD removal dropped to 40% from 90% and total selenium removal dropped to 0% from 60% in the period V. When the selenite feed was stopped in period VI, the COD removal did not improve suggesting irreversible crashing of reactor. The total selenium removed per unit of biomass in the aqueous phase was 33.98 mg Se gTSS^{-1} and 36.65 mg of Se gTSS^{-1} when operating the continuous reactor at TSS concentrations of 1300 (period II; Figure 9.3) and 3000 mg L^{-1} (period V; Figure S2a in Appendix 4), respectively.

9.4. Discussion

9.4.1. Entrapment of BioSeNPs in the activated sludge

This is the first study demonstrating selenite reduction by activated sludge and trapping of the produced BioSeNPs in the activated sludge flocs (Figure 9.2c, 9.2d, 9.3c). The reduction of the total selenium concentration in the supernatant suggests that the BioSeNPs are either trapped in the biomass or volatilized and subsequently trapped in gas traps (Figure 9.2c, 9.2d, 9.3b, 9.4). The negligible presence of selenium in these gas traps and the significant presence of selenium in the biomass (Figure 9.3c) suggest that the majority of the produced BioSeNPs are trapped in the activated sludge flocs.

Lower temperatures and higher DO levels led to the trapping of BioSeNPs in the activated sludge flocs (Figures 9.2c, 9.3b, 9.3c). At lower DO levels, anoxic and anaerobic zones might be present, where dissimilatory reduction of selenite could take place extracellularly or in the periplasm resulting in extracellular synthesis of BioSeNPs (Figure 9.3b, 9.3c) (Li et al., 2014). Selenite reduction at higher temperatures (45 °C in this study) had a lower dissolved oxygen concentration in the shake flasks (data not shown), which might have also allowed BioSeNPs production extracellularly, resulting in higher total selenium concentrations in the medium (Figure 9.2c).

The two different mechanisms were also evident in the total selenium removal kinetics at different DO levels (Figure 9.3b). The total selenium concentration in the supernatant was consistently higher at low DO level than the one observed at high DO level, even when the dissolved selenium concentrations were similar (Figure 9.3b). This suggests that reduction of selenite in the activated sludge studied is due to a detoxification mechanism at high DO levels and not via dissimilatory reduction (Dhanjal and Cameotra, 2010). The detoxification mechanism in this study involves the reduction of selenite to BioSeNPs rather than the conversion to volatilized species, indeed only minor quantities of selenium were trapped in gas phase washing bottles (Figure 9.3c). The presence of different reduction mechanisms by a non-specialized biomass suggests the development of a new population depending on the DO concentration, as was indeed observed in the DGGE analysis (Figure 9.5).

9.4.2. Practical implications

Glucose was clearly the more favorable substrate for selenite reduction by activated sludge compared to lactate and acetate (Figure 9.2d), due to the higher amount of electrons donated by glucose while converting to CO_2. A similar preference by activated sludge during the reduction of Cr(VI) was observed when glucose, lactose, cheese whey, acetate and citrate were used as carbon sources (Ferro Orozco et al., 2010).

It is interesting to note that no growth in the TSS of the biomass fed with sodium selenite (172 mg L^{-1} or 1mM) was observed. Similarly, very low biomass growth (less than 70% as compared to control) was observed when selenite was reduced under aerobic conditions by *Phanerochaete chrysosporium* (Espinosa-Ortiz et al., 2014). Similar non-growth related detoxification mechanisms have been reported for the microbial decolorization of reactive red 22 by *Pseudomonas luteola*, a toxic dye used in textile and paper industry (Chen, 2002).

The crashing and non-recoverability of the continuous reactor at two different TSS concentrations (Figure 9.4 and Figure S2a in Appendix 4) suggest that the toxicity of selenite was irreversible. On the other hand, anaerobic reduction of selenate and selenite in a UASB reactor may remove similar loading rates without a apparent toxicity but it would lead to presence of BioSeNPs in the effluent. Sequencing batch feeding can be used to overcome the toxicity of selenite, as has been used for aerobic reduction of Cr(VI) to Cr(III) (Dermou et al., 2005).

The higher DO requirement for trapping elemental selenium in the activated sludge flocs would entail higher cost for selenite reduction under aerobic conditions. This high cost can be offset by the use of the activated sludge trapping selenium as a slow selenium release fertilizer, which might requires no further processing (Haug et al., 2007). The use of an alternative electron donor, such as cheese whey (Ferro Orozco et al., 2010), could further reduce the treatment costs without affecting the performance. Furthermore, the successful implementation of such a process would lead to a single step selenium removal and recovery process vis-a-vis the two-step anaerobic selenium removal, where high recovery efficiencies can only be achieved by applying a chemical (Buchs et al., 2013) or electrocoagulation (Staicu et al., 2014) post-treatment step.

9.5. Conclusions

This study provided the proof-of-concept for the removal of total selenium fed as selenite from the aqueous phase in a one-step process. The higher DO levels and lower temperature were required to ensure trapping of produced BioSeNPs in the activated sludge flocs. This study also showed that the glucose was the best carbon source to achieve maximum selenium removal per g of COD. The DGGE analysis showed the possibility of rise of new microbial populations. The crashing of continuous reactor at two different TSS values suggests the need for different strategy for continuous reactor operations.

Acknowledgements

The authors are thankful to Ferdi Battles, Lyzette Robbemont, Berend Lolkema (UNESCO-IHE, Delft) for the ICP-MS measurements, and Marina Seder-Colomina (Université Paris-Est, Marne la Vallée), and Julie Cosmidis (UPMC Univ Paris 06, Paris) for TEM-EDXS measurements.

9.6 References

Ahammad, S.Z., Davenport, Read, L.F., Gomes, J., Sreekrishnan, T.R., Dolfing, J., 2013. Rational immobilization of methanogens in high cell density bioreactors. RSC Adv. 3, 774–781.

APHA, 2005. Standard methods for examination of water and wastewater, 5th ed. American Public Health Association, Washington, DC, USA.

Buchs, B., Evangelou, M.W.-H., Winkel, L., Lenz, M., 2013. Colloidal properties of nanoparticular biogenic selenium govern environmental fate and bioremediation effectiveness. Environ. Sci. Technol. 47, 2401–2407. s

Cantafio, A.W., Hagen, K.D., Lewis, G.E., Bledsoe, T.L., Nunan, K.M., Macy, J.M., 1996. Pilot-Scale Selenium Bioremediation of San Joaquin Drainage Water with Thauera selenatis. Appl. Environ. Microbiol. 62, 3298–303.

Chen, B., 2002. Understanding decolorization characteristics of reactive azo dyes by Pseudomonas luteola : toxicity and kinetics. Process Biochem. 38, 437–446.

Cosmidis, J., Benzerara, K., Menguy, N., Arning, E., 2013. Microscopy evidence of bacterial microfossils in phosphorite crusts of the Peruvian shelf: Implications for phosphogenesis mechanisms. Chem. Geol. 359, 10–22.

Dermou, E., Velissariou, A., Xenos, D., Vayenas, D. V, 2005. Biological chromium(VI) reduction using a trickling filter. J. Hazard. Mater. 126, 78–85.

Dhanjal, S., Cameotra, S.S., 2010. Aerobic biogenesis of selenium nanospheres by Bacillus cereus isolated from coalmine soil. Microb. Cell Fact. 9, 52.

Espinosa-Ortiz, E.J., Gonzalez-Gil, G., Saikaly, P.E., van Hullebusch, E.D., Lens, P.N.L., 2014. Effects of selenium oxyanions on the white-rot fungus Phanerochaete chrysosporium. Appl. Microbiol. Biotechnol. (accepted) doi:10.1007/s00253-014-6127-3

Ferro Orozco, A.M., Contreras, E.M., Zaritzky, N.E., 2010. Cr(Vi) reduction capacity of activated sludge as affected by nitrogen and carbon sources, microbial acclimation and cell multiplication. J. Hazard. Mater. 176, 657–65.

Haug, A., Graham, R.D., Christophersen, O.A., Lyons, G.H., 2007. How to use the world's scarce selenium resources efficiently to increase the selenium concentration in food. Microb. Ecol. Health Dis. 19, 209–228.

Jain, R.; Gonzalez-Gil, G.; Singh, V., van Hullebusch, E.D., Farges, F.; Lens, P.N.L., 2014. Biogenic selenium nanoparticles, Production, characterization and challenges. In Kumar, A., Govil, J.N., Eds. Nanobiotechnology. Studium Press LLC, USA, pp. 361-390

Jain, R., Jordan, N., Schild, D., Hullebusch, E.D. Van, Weiss, S., Franzen, C., Hubner, R., Farges, F., Lens, P.N.L., 2015. Adsorption of zinc by biogenic elemental selenium nanoparticles. Chem. Eng. J. 260, 850–863.

Johnson, N.C., Manchester, S., Sarin, L., Gao, Y., Kulaots, I., Hurt, R.H., 2008. Mercury vapor release from broken compact fluorescent lamps and in situ capture by new nanomaterial sorbents. Environ. Sci. Technol. 42, 5772–5778.

Karya, N.G.A., van der Steen, N.P., Lens, P.N.L., 2013. Photo-oxygenation to support nitrification in an algal-bacterial consortium treating artificial wastewater. Bioresour. Technol. 134, 244–50.

Lenz, M., Hullebusch, E.D. Van, Hommes, G., Corvini, P.F.X., Lens, P.N.L., 2008. Selenate removal in methanogenic and sulfate-reducing upflow anaerobic sludge bed reactors. Water Res. 42, 2184–2194.

Lenz, M., Lens, P.N.L., 2009. The essential toxin: the changing perception of selenium in environmental sciences. Sci. Total Environ. 407, 3620–3633.

Li, D.-B., Cheng, Y.-Y., Wu, C., Li, W.-W., Li, N., Yang, Z.-C., Tong, Z.-H., Yu, H.-Q., 2014. Selenite reduction by *Shewanella oneidensis* MR-1 is mediated by fumarate reductase in periplasm. Sci. Rep. 4, 3735. 5

Muyzer, G., de Waal, E., Uitterlinden, A.G., 1993. Profiling of complex microbial populations by denaturing gradient gel electrophoresis analysis of polymerase chain. Appl. Environ. Microbiol. 59, 695–700.

Staicu, L.C., van Hullebusch, E.D., Lens, P.N.L., Pilon-Smits, E.A., Oturan, M. a, 2014. Electrocoagulation of colloidal biogenic selenium. Environ. Sci. Pollut. Res. Int. (accepted) doi:10.1007/s11356-014-3592-2

Tejo Prakash, N., Sharma, N., Prakash, R., Raina, K.K., Fellowes, J., Pearce, C.I., Lloyd, J.R., Pattrick, R. a D., 2009. Aerobic microbial manufacture of nanoscale selenium: exploiting nature's bio-nanomineralization potential. Biotechnol. Lett. 31, 1857–1862.

Winkel, L., Feldmann, J., Meharg, A.A., 2010. Quantitative and qualitative trapping of volatile methylated selenium species entrained through nitric acid. Environ. Sci. Technol. 44, 382–387.

CHAPTER 10

Entrapped elemental selenium nanoparticles increases settleablity and hydrophilicity of activated sludge

This chapter is accepted for publication in modified form:

Jain, R., Seder-Colomina, M. Jordan, N., Cosmidis, J., Hullebusch, E.D. Van, Weiss, S., Dessi, P., Farges, F., Lens, P.N.L., 2015. Entrapped elemental selenium nanoparticles increases settleablity and hydrophilicity of activated sludge. J. Hazad. Mat. 295, 193-200.

Abstract:

Selenium containing wastewaters can be treated in activated sludge systems, where the selenium oxyanions are reduced to elemental selenium by activated sludge, the latter is entrapped in the activated sludge flocs. So far, no studies have been carried out on the characterization of selenium fed activated sludge flocs, which is important for the development of this process. This study showed that more than 93% of the trapped selenium in activated sludge flocs is in the form of elemental selenium, both as amorphous selenium nanospheres and trigonal selenium nanorods. The entrapment of the elemental selenium nanoparticles affects the physico-chemical properties of activated sludge. Selenium fed activated sludge has faster settling rates and lower hydrophobicity compared to the control activated sludge. The selenium fed activated sludge showed less negative surface charge density, most likely due to the presence of elemental selenium nanoparticles. However, selenium fed activated sludge has a poorer dewaterability at higher total suspended solids concentrations as compared to control activated sludge.

Keywords: selenium nanoparticles, activated sludge, settleability, hydrophobicity, surface charge

Graphical abstract:

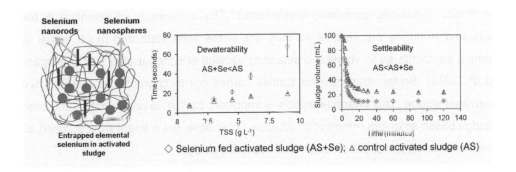

◇ Selenium fed activated sludge (AS+Se); △ control activated sludge (AS)

10.1. Introduction

Selenium is an essential element and is required in low doses for synthesizing selenoproteins, preventing cardiovascular disease, assisting in sperm mobility and avoiding miscarriage (Rayman, 2000). However, at higher doses selenium oxyanions (selenate and selenite) can cause toxicity to humans, but also to animals and aquatic organisms. Thus, they need to be removed from wastewaters prior to discharge (Lenz and Lens, 2009; Winkel et al., 2012; Wu, 2004). Physico-chemical remediation is expensive and yet sometimes ineffective in achieving the stringent selenium discharge criteria (less than 50 µg L^{-1}) (Lenz and Lens, 2009). Anaerobic microbial reduction of selenium oxyanions to elemental selenium is often a recommended biological process for selenium oxyanions containing wastewater treatment (Cantafio et al., 1996; Lenz et al., 2008). However, biological anaerobic reduction of selenium oxyanions leads to the presence of colloidal elemental selenium nanoparticles (BioSeNPs) in the discharged wastewaters (Jain et al., 2014a). These BioSeNPs have to be removed prior to discharge, which requires an additional treatment step that further increases the remediation cost (Buchs et al., 2013).

Aerobic reduction of selenium oxyanions to either volatilized selenium compounds followed by gas trapping (Kagami et al., 2013) or selenium trapped in biomass such as *Bacillus cereus* (Dhanjal and Cameotra, 2010), *Escherichia coli* (Dobias et al., 2011) or activated sludge (Jain et al., 2014) would be a one-step process for treatment of selenite containing wastewaters. The entrapment of selenium in the activated sludge is progressive as they are in the solid state and much easier to handle as compared to volatilized selenium trapped in concentrated HNO$_3$ (Kagami et al., 2013). So far, there are no studies carried out to characterize selenium fed activated sludge. This characterization is important for the development of activated sludge based selenium remediation process. Therefore, the present work focused on the characterization of the selenium fed activated sludge.

In this study, activated sludge loaded with selenium was produced by aerobic reduction of selenite. The localization, speciation and crystallinity of selenium in the activated sludge flocs were identified by Scanning Electron Microscopy-Energy Disperse X-ray Spectroscopy (SEM-EDXS), Transmission Electron Microscopy

(TEM), sequential extraction, X-ray diffraction (XRD) and selected area electron diffraction (SAED). The elemental constituents, morphology, hydrophobicity, sludge volume index (SVI), capillary suction time (CST) and surface groups density of selenium fed activated sludge were compared with control activated sludge without selenium.

10.2. Materials and methods

10.2.1. Production of activated sludge loaded with selenium and control activated sludge

Activated sludge used in this study was collected from a full scale domestic wastewater treatment plant in Harnaschpolder (The Netherlands). The synthetic wastewater was composed of NH_4Cl (600 mg L^{-1}), $MgCl_2.2H_2O$ (200 mg L^{-1}), $CaCl_2.2H_2O$ (20 mg L^{-1}). Na_2SeO_3 (173 mg L^{-1}) and glucose (2000 mg L^{-1}) were used for selenium and carbon source, respectively. The total suspended solid (TSS) concentration added was 1300 ± 100 mg L^{-1}. Mixing was carried out with a magnetic stirrer (500 rpm). The selenite reduction was carried out under continuous aeration by an air flow (15-20 L air h^{-1}). After the reduction of selenite, selenium fed activated sludge was collected by settling and the supernatant was discarded. Control activated sludge without the addition of sodium selenite was produced using the same constituents and collected by settling. The pH was maintained at 7.5-8.0 by manual addition of Na_2CO_3 when needed.

10.2.2. Elemental determination in the selenium fed activated sludge

10 mL of concentrated HNO_3 was added to 0.5 g (TSS) of selenium fed as well as control activated sludge. The sludge was destroyed using microwave digestion by heating to 175 °C for 45 minutes. The samples were appropriately diluted and elemental concentrations were measured by induced couple plasma mass spectroscopy (ICP-MS). The experiments were done in triplicate.

10.2.3. Selenium localization and speciation determination

SEM-EDXS and TEM were carried out to locate selenium in the microorganisms and activated sludge flocs. For SEM-EDXS analysis, each sample was diluted in Milli-Q water and filtered using 0.22 μm pore-size polycarbonate membrane filters. Filters were deposited on carbon tape, dried at ambient temperature and finally coated with a thin carbon film. Samples for TEM were diluted in Milli-Q water, deposited copper TEM grid covered with a lacey formvar film and dried at ambient temperature.

Selenium species concentration was determined by carrying out sequential extraction following the protocol published by Wright et al. (Wright et al., 2003) but using a 30 times higher extractant to solid ratio for complete selenium recovery. Sequential extraction analysis was carried out in quadruplicates. The crystallinity of the trapped selenium was determined by XRD and SAED associated with the TEM.

10.2.4. Physico-chemical properties of activated sludge

The SVI and CST of the selenium fed activated sludge and control activated sludge were determined as per standard methods (APHA, 2005). For SVI, 1.8 g L^{-1} of TSS was used. The relative hydrophobicity (RH) was measured following the protocol described in Laurent et al. (2009a). Fourier Transform-Infrared spectroscopy (FT-IR) was carried out to determine the functional groups present in the sludge.

Acid-base titration was carried out for determining the pK_a of the functional groups using a Metrohm autotitrator unit. 0.0266 g of selenium fed and control activated sludge (TSS) was suspended in 30 mL of Milli-Q water with 1 mM of NaCl background electrolyte. The titration was carried by automatic addition of 0.1 mL HCl (0.01214 M). The acid-base titration data were fitted using a PROTOFIT software (Turner and Fein, 2006) as described by Laurent et al. (2009b). Briefly, a non-electrostatic model with four discrete acidic sites and an extended Debye-Huckel activity coefficient were used. Prior to simulation using PROTOFIT, the derivative of the acid-base titration versus moles of HCl added was plotted to determine the local minima. These local minima represent the pK_a of the functional groups (Braissant et al., 2007). Taking these pK_a as the initial guess, the data fitting was carried out.

10.2.5. Analytical methods

Selenium measurements by ICP-MS and XRD analysis were carried out as described in previous studies (Jain et al., 2015). For IR spectroscopy, KBr pellets were prepared by mixing approximately 1 mg of the samples with 300 mg dried KBr. Clear pellets were obtained after pressing for 2 minutes at 145,000 psi. A Bruker Vertex 70/v spectrometer equipped with a D-LaTGS-detector (L-alanine doped triglycine sulfate) was used to measure FT-IR spectra of the sludge samples. The measurements were carried over the range 4000-400 cm^{-1} in the transmittance mode, with a spectral resolution of 4 cm^{-1}. Each spectrum was averaged out over 64 scans.

SEM observations were performed with a Field Emission Gun Scanning Electron Microscope (GEMINI ZEISS Ultra55) operated at 2 kV. SEM-EDXS analyses were performed on the selected particles at 15 kV. TEM and SAED analyses were performed using a JEOL 2100F (FEG) operating at 200 kV and equipped with a field emission gun, a high-resolution UHR pole piece, and a Gatan energy filter GIF 200.

10.3. Results

10.3.1. Concentration of various elements in selenium fed activated sludge

The selenite was reduced to red elemental selenium as observed from the appearance of the red color in the activated sludge. More than 78% of the fed selenium was entrapped in the biomass. The concentration of trapped selenium in the activated sludge is 54.9 ± 2.3 mg of Se per g of TSS. The concentrations of Na, Mg Al, K, Ca and Fe in selenium fed activated sludge are presented in Table 10.1. Other elements such as Cu, Mn, Zn, Ba, Cr, V, Ni, Co, Pb and Mo were less than 1.0 mg per g of TSS.

Table 10.1. Concentrations of various common elements in the selenium fed (As+Se).

Se	Na	Mg	Al	K	Ca	Fe
mg of elements per g of TSS						

AS+Se	54.9 ± 2.3	26.9 ± 4.3	4.6 ± 0.3	3.1 ± 0.7	2.9 ± 1.8	31.4 ± 1.6	14.0 ± 0.3

10.3.2. Characterization of trapped selenium in the activated sludge

Sequential extraction of selenium trapped in the activated sludge suggests that the 93.3 ± 9.7% of trapped selenium is in the form of elemental selenium. The sequential extraction method was able to account for 95.7% of the total selenium trapped in the activated sludge. All other fractions such as soluble/exchangeable fraction, adsorbed fraction and residual fraction were insignificant as they constitute only 2.4 ± 0.6% of the total trapped selenium.

SEM images showed the presence of two different morphologies of selenium: nanospheres and nanorods (Figure 10.1a). EDX spectra confirmed that these morphologies are composed of selenium (Figure 10.1b1, 10.1b2). The presence of selenium nanorods and nanospheres was further confirmed in SEM images and its corresponding cartography (Figure 10.1.e, f). SEM image of the bacterial cells suggests that the some fractions of the selenium trapped in the biomass are present inside the bacterial cells (Figure 10.1c). The white colored spheres in the Figure 10.1c are selenium nanospheres. The black dots in the Figure 10.1c are pores of the filters. The TEM image showed that the selenium nanospheres are closely associated with bacterial cells. Some of the spheres are indeed present at the surface of the bacteria, while others are possibly found inside the bacteria, confirming SEM observations (Figure 10.1d).

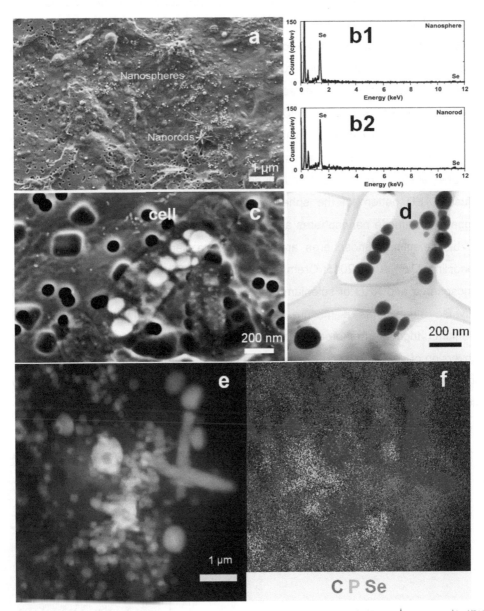

Figure 10.1. SEM (a) and EDX spectra of nanospheres (b1) and nanorods (b2) present in the selenium fed activated sludge. Zoomed in SEM image (c), TEM image (d) suggests the presence of selenium inside the cells and flocs and (e) SEM images and corresponding cartography (f) clearing showing presence of selenium nanorods.

Biogenic elemental selenium exist mainly in trigonal, monoclinic or amorphous forms (Jain et al., 2014a). In order to determine the crystallinity of trapped elemental selenium, XRD was carried out on selenium fed activated sludge. XRD results showed the features at 23.4 and 30.0 2-theta values corresponding to 100 and 101 facet of trigonal selenium, respectively while the feature at 26.9 2-theta value corresponds to 220 facet of β-monoclinic selenium (Ding et al., 2002) (Figure 10.2a). The XRD also showed a hump like structure between 15 to 40 2-theta values suggesting the presence of amorphous structures. In contrast to selenium fed activated sludge, control activated sludge showed no features in the XRD. The diffuse SAED pattern of the spherical particles present in the activated sludge suggests that these nanospheres are either amorphous or poorly crystalline (Figure 10.2b). Selenium nanospheres are known to be generally either amorphous or monoclinic (Jain et al., 2015; Oremland et al., 2004). Thus, the SAED pattern of the nanospheres and XRD of selenium fed activated sludge suggest that selenium fed activated sludge has a mixture of different crystalline structure of elemental selenium: trigonal (nanorods) and monoclinic (nanospheres). The biological conversion of selenium oxyanions always results in the formation of amorphous or β-monoclinic nanospheres (Oremland et al., 2004; Pearce et al., 2009), which may later transform to trigonal nanorods (Zhang et al., 2011).

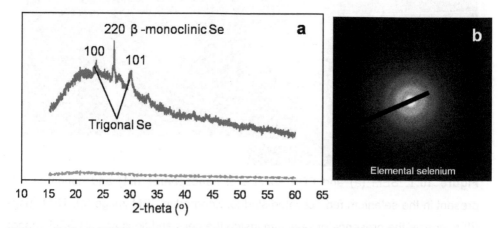

Figure 10.2. (a) XRD of selenium fed (—) and control (without selenium, —·—·-) activated sludge and (b) SAED pattern of trapped elemental selenium nanospheres. 3.3 Spatial distribution of elements in activated sludge flocs

Elemental selenium is known to adsorb heavy metals including mercury (Fellowes et al., 2011; Jiang et al., 2012; Johnson et al., 2008), copper (Bai et al., 2011) and zinc (Jain et al., 2015). Figure 10.3a shows the SEM image of a single bacterium from selenium fed activated sludge. Figure 10.3b showed the corresponding EDXS analysis, which indicates that the nanospheres observed in the SEM image of Figure 10.3a are composed of selenium. Al and Mg were found to be closely associated to the selenium particles (Figures 10.3c, d). Other elements such as Ca, Fe, Cu, Zn, Pb and Ba were not found to have gradient towards elemental selenium trapped in the sludge (data not shown). It is important to note that the distortion of one of the nanosphere of selenium observed in Figure 10.3a is due to the damage by the electron beam while making the measurements.

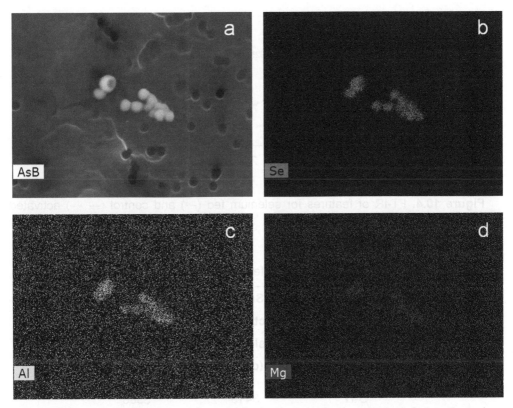

Figure 10.3. (a) SEM image of a single bacterium from selenium fed activated sludge and EDXS analyses corresponding to the spatial distribution of (b) Se, (c) Al and (d) Mg in the SEM image.

10.3.4. Functional groups in the activated sludges

FT-IR analysis of the selenium fed activated sludge and the control sludge showed the presence of hydroxyl (3419-3386 cm⁻¹) (Wang et al., 2011), aliphatic hydrocarbons (2958-2925 cm⁻¹) (Wang et al., 2012), amide (1652, 1540, 1236 cm⁻¹) (Wang et al., 2012) and polysaccharides (1074-1056 cm⁻¹) groups (Xu et al., 2011) (Figure 10.4). The details of the FT-IR for both selenium fed and control activated sludge are captured in Table 10.2. Selenium is known to show no features at wavenumbers 4000 to 800 cm⁻¹ (Nakamura and Ikawa, 2001). The FT-IR confirms the presence of carboxyl and amine groups. A weak shoulder at 1161 and 1151 cm⁻¹ suggests the occurrence of phosphate groups.

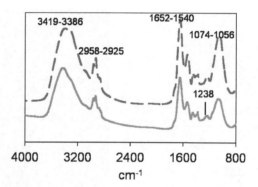

Figure 10.4. FT-IR of features for selenium fed (—) and control (— —) activated sludge.

Table 10.2. FT-IR features of selenium fed and control activated sludges.

Functional groups	Se fed activated sludge (cm⁻¹)	Control activated sludge (cm⁻¹)
O-H	3419	3386
C-H	2958-2925	2958-2925
C=O	1652	1652

C-N, N-H	1540	1540
CH$_2$	1456	1454
COO	1378	1378
C-N, N-H	1238	1236
P=O	1161	1156
C-O, C-O-C	1074	1056

To further evaluate the surface charge present on the selenium fed activated sludge, an acid-base titration was carried out. The delta pH versus micro-moles of HCl added was plotted in Figure 10.5. The local maxima in the Figure 10.5a represent the maximum shift in pH and hence the equivalence points. Similarly, the local minima represent the minimum change in pH and hence the pK$_a$ of the functional groups present (Braissant et al., 2007). The local minima for selenium fed activated sludge are observed at pH 7.1, 6.9, 5.2, 3.7 and 3.4 (Figure 10.5a). Similarly, the local minima for control activated sludge are observed at pH 7.2, 5.1 and 3.2. The local minima at 6.9 - 7.2 correspond to either sulfinic acids, sulfonic acids or thiol groups (Braissant et al., 2007). Carboxyl or phosphoryl groups can be assigned to local minima at pH 5.1-5.2 (Martinez et al., 2002). The local minima observed at pH of 3.2-3.7 are due to the presence of carboxyl groups (Martinez et al., 2002).

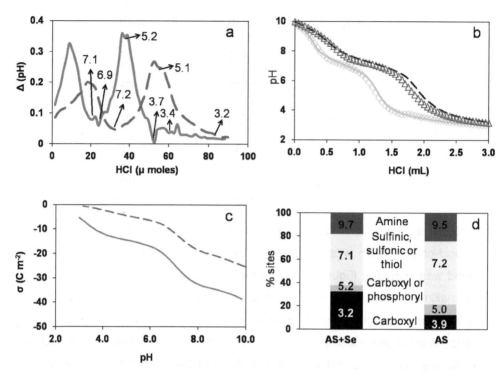

Figure 10.5. (a) Derivative of acid-base titration data of selenium fed (—) and control (— —) activated sludge, (b) Titration data for selenium fed (◊ Raw, — Simulated) and control (Δ Raw, — — Simulated) activated sludge, (c) Surface charge density of activated sludge trapping selenium (—) and control activated sludge (— —) and (d) % sites with corresponding pKₐ values of various functional groups for selenium fed and control activated sludge.

The simulation of the acid-base titration data fitted well with the experimental data (Figure 10.5b). The surface charge density of the selenium fed activated sludge was two times more negative than the control activated sludge at neutral pH (Figure 10.5c). The simulation of acid-base titration data predicted the pKₐ of functional groups at pH of 3.2, 5.2, 7.1 and 9.7 for selenium fed activated sludge (Figure 10.5d). For the control activated sludge, predicted pKₐ values of functional groups are at pH 3.9, 5.0, 7.2 and 9.5. It is important to note that we may not observe the local minima in the derivate of pH vs moles of acid graph at the start and end of the titration due to a small change in pH and hence the most acidic and basic groups may be missed out. But as shown above, these groups can be successfully predicted by the simulation.

10.3.5. Physical properties of selenium fed and control activated sludge

The SVI of the selenium fed and control activated sludge was 61.1 ± 0.3 and 138.8 ± 0.1 mL g^{-1} (Figure 10.6a). The CST of the selenium fed activated sludge and control activated sludge was in the range of 20 seconds at a TSS of 3 g L^{-1} (Figure 10.6b). With the increase in TSS, the CST of the selenium fed activated sludge increased to a high value of 67.2 s as compared to 19.0 s for control activated sludge at TSS value of 9 g L^{-1}. The relative hydrophobicity of selenium fed activated sludge was 1.6 times lower than that of the control activated sludge (Figure 10.6c).

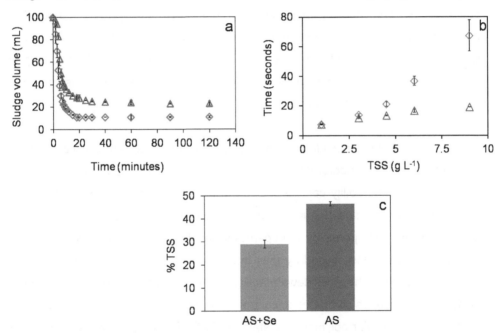

Figure 10.6. SVI (a), CST (b) and RH (c) of selenium fed (◊) and control (Δ) activated sludge.

10.4. Discussion

10.4.1 Entrapment of elemental selenium affects the physical properties of activated sludge flocs

This study demonstrated for the first time the trapping of microbiologically produced inorganic particles, i.e. elemental selenium, in activated sludge flocs improves their settleability, decreases their relative hydrophobicity, but negatively impacts their dewaterability. Sequential extraction experiments suggest that the elemental form of the selenium is the main constituent of the trapped selenium. Elemental selenium is 4.5 times denser than the activated sludge, thus the improved settleability can be attributed to the dense elemental selenium nanoparticles that increase the density of activated sludge and thus, lead to better settling. The lower RH of activated sludge with entrapped elemental selenium as compared to control activated sludge suggests the presence of more polar groups such as hydroxyl, carboxyl or phosphoryl groups in the flocs (Laurent et al., 2009a). The larger presence of carboxyl groups on the surface was verified by higher percentage of carboxyl sites and less negative surface charge in the selenium fed activated sludge as compared to control activated sludge (Figure 10.5c, d).

The higher CST of selenium fed activated sludge at high TSS values as compared to the control indicates a poorer dewaterability of the sludge (Figure 10.6b). The reason for the poor dewaterability can be due to the blockage of filter pores by elemental selenium (nano)particles (Figure 10.1c, d), thus, leading to a higher CST. The blocking of the filter pores by nanoparticles can be due to the agglomeration of selenium nanoparticles caused by their interaction with metals such as Al and Mg (Jain et al., 2015). The poor dewaterability of the activated sludge with entrapped elemental selenium as compared to control activated sludge can also be attributed to a different extracellular polymeric substances (EPS) content (Li and Yang, 2007; Ye et al., 2011). Selenite reduction has been shown to cause stress on *Bacillus* sp., leading to the production of larger amounts of EPS, with a different composition, as compared non stressed microorganisms (Xu et al., 2009). It is important to note that the entrapped selenium does not affect the CST at TSS concentrations (< 3 g L^{-1}) normally maintained in activated sludge reactors.

10.4.2 Entrapment of elemental selenium affects the surface properties of the activated sludge flocs

The less negative surface charge density of the selenium fed activated sludge flocs as compared to the control activated sludge is due to the higher site concentration of carboxyl groups as well as their lower pK_a value (3.2 for selenium fed activated sludge as compared to 3.9 for control activated sludge). The difference in the EPS content can be a one of the reason for the difference in the surface charge density of the selenium fed and control activated sludge. The other reason for less negative selenium fed activated sludge is the presence of elemental selenium nanoparticles. Elemental selenium nanoparticles are known to have a negative ζ-potential with an acidic iso-electric point (pH 3.5) due to the presence of organic moieties produced by microorganisms on their surface (Buchs et al., 2013; Dhanjal and Cameotra, 2010). These negative ζ-potential nanoparticles might contribute to the less negative surface density of the selenium fed activated sludge.

Due to the negative ζ-potential of these elemental selenium nanoparticles, the spatial distribution showed a higher concentration of Al and to some extent of Mg near the elemental selenium nanoparticles trapped in the activated sludge flocs (Figure 10.3), as observed in case of Cu (Bai et al., 2011), Zn (Jain et al., 2015) as well as other cations such as Ca and Ba (Buchs et al., 2013). However, it is interesting to see that other than Al and Mg, all the other possible elements such as Ca, Ba, Fe, Na did not show any concentration gradient towards elemental selenium.

The preference of any cation towards an adsorbent follows 1) a decrease in ionic radius, 2) an increase in electronegativity of the metal ion or 3) an increase in ratio of the ionisation potential and ionic radius (McKay and Porter, 1997). The ionic radius of Al was the smallest among the metals tested (Table 10.1) other than Mn. However, the concentration of Mn was 15 times lower than that of Al and thus Al would have outcompeted Mn for the sites on the elemental selenium nanoparticles. Moreover, as Al is a trivalent ion as compared to divalent Mn, the tendency of Al for forming complexes with the protein coatings of EPS present on elemental selenium (Jain et al., 2014) would be higher than Mn and thus would also outcompete Mn. The case of Mg being the second most for the preference of elemental selenium also follows the smaller ionic radius theory. However, Fe has a smaller ionic radius than Mg (0.645 Å as compared to 0.72 Å), higher electronegativity (1.88 compared to 1.31), higher ratio of ionization potential to ionic radius (-0.68 compared to -3.3) and also a higher concentration (14.0 Fe as compared to 4.6 mg of mg per g of TSS). However, Fe(III) has a very low solubility and may be forming phosphates or oxides and hence non-available to compete for adsorption sites on elemental selenium nanoparticles. For similar reason, Fe would not compete with Al for adsorption sites on elemental selenium nanoparticles. Ca has higher ionic radius than Mg but it is almost 9 times higher in concentration than Mg and thus would have shown preference for elemental selenium nanoparticles as observed in Jain et al. (Jain et al., 2015). However, the absence of Ca gradient towards elemental selenium nanoparticles also suggests that it was not available for binding to elemental selenium nanoparticles in the activated sludge.

10.5. Conclusions

This study showed that most of the trapped selenium in activated sludge flocs is in the form of elemental selenium. SEM and TEM images suggest that elemental selenium is present both inside the flocs as well as inside the microbial cell. XRD and SAED suggest the presence of amorphous and β-monoclinic nanospheres and trigonal nanorods. The trapping of this elemental selenium in the activated sludge flocs improves the settleability and lowers the hydrophobicity of the sludge. However, the presence of the elemental selenium affects the dewaterability of the activated sludge, but only at higher TSS concentrations. The presence of selenium nanoparticles led to Al and Mg concentrations gradient in the selenium fed activated sludge.

Acknowledgements
The authors would like to acknowledge Ferdi Battles (UNESCO-IHE, The Netherlands) for ICP-MS analysis, Harald Foerstendorf and Karsten Heim (HZDR, Germany) for FT-IR analysis and Rohit Kacker (TU Delft, The Netherlands) for XRD analysis.

10.6. References

APHA, 2005. Standard methods for examination of water and wastewater, 5th ed. American Public Health Association, Washington, DC, USA.

Bai, Y., Rong, F., Wang, H., Zhou, Y., Xie, X., Teng, J., 2011. Removal of copper from aqueous solution by adsorption on elemental selenium nanoparticles. J. Chem. Eng. Data 56, 2563–2568.

Braissant, O., Decho, a. W., Dupraz, C., Glunk, C., Przekop, K.M., Visscher, P.T., 2007. Exopolymeric substances of sulfate-reducing bacteria: Interactions with calcium at alkaline pH and implication for formation of carbonate minerals. Geobiology 5, 401–411.

Buchs, B., Evangelou, M.W.-H., Winkel, L., Lenz, M., 2013. Colloidal properties of nanoparticular biogenic selenium govern environmental fate and bioremediation effectiveness. Environ. Sci. Technol. 47, 2401–2407.

Cantafio, A.W., Hagen, K.D., Lewis, G.E., Bledsoe, T.L., Nunan, K.M., Macy, J.M., 1996. Pilot-Scale Selenium Bioremediation of San Joaquin Drainage Water with Thauera selenatis. Appl. Environ. Microbiol. 62, 3298–303.

Dhanjal, S., Cameotra, S.S., 2010. Aerobic biogenesis of selenium nanospheres by Bacillus cereus isolated from coalmine soil. Microb. Cell Fact. 9, 52.

Ding, Y., Li, Q., Jia, Y., Chen, L., Xing, J., Qian, Y., 2002. Growth of single crystal selenium with different morphologies via a solvothermal method. J. Cryst. Growth 241, 489–497.

Dobias, J., Suvorova, E.., Bernier-latmani, R., 2011. Role of proteins in controlling selenium nanoparticle size. Nanotechnology 22, 195605.

Fellowes, J.W., Pattrick, R.A.D., Green, D.I., Dent, A., Lloyd, J.R., Pearce, C.I., 2011. Use of biogenic and abiotic elemental selenium nanospheres to sequester elemental mercury released from mercury contaminated museum specimens. J. Hazard. Mater. 189, 660–669.

Jain,R., Gonzalez-Gil, G., Singh, V., van Hullebuchs, Eric, D., Farges, F., Lens, P.N.L., 2014. Biogenic selenium nanoparticles : production , characterization and challenges, in: Kumar, A., Govil, J, N. (Eds.), Nanobiotechnology. Studium Press LLC, USA, pp. 361–390.

Jain, R., Jordan, N., Schild, D., Hullebusch, E.D. Van, Weiss, S., Franzen, C., Hubner, R., Farges, F., Lens, P.N.L., 2015. Adsorption of zinc by biogenic elemental selenium nanoparticles. Chem. Eng. J. 260, 850–863.

Jain, R., Matassa, S., Singh, S., van Hullebusch, E.D., Esposito, G., Lens, P.N.L., 2014. reduction to elemental selenium nanoparticles by activated sludge under aerobic conditions. Biochem. Eng. J (submitted, Chapter 9).

Jiang, S., Ho, C.T., Lee, J.-H., Duong, H. Van, Han, S., Hur, H.-G., 2012. Mercury capture into biogenic amorphous selenium nanospheres produced by mercury resistant Shewanella putrefaciens 200. Chemosphere 87, 621–624.

Johnson, N.C., Manchester, S., Sarin, L., Gao, Y., Kulaots, I., Hurt, R.H., 2008. Mercury vapor release from broken compact fluorescent lamps and in situ capture by new nanomaterial sorbents. Environ. Sci. Technol. 42, 5772–5778.

Kagami, T., Narita, T., Kuroda, M., Notaguchi, E., Yamashita, M., Sei, K., Soda, S., Ike, M., 2013. Effective selenium volatilization under aerobic conditions and recovery from the aqueous phase by Pseudomonas stutzeri NT-I. Water Res. 47, 1361–1368.

Laurent, J., Casellas, M., Dagot, C., 2009a. Heavy metals uptake by sonicated activated sludge: relation with floc surface properties. J. Hazard. Mater. 162, 652–660.

Laurent, J., Casellas, M., Pons, M.N., Dagot, C., 2009b. Flocs surface functionality assessment of sonicated activated sludge in relation with physico-chemical properties. Ultrason. Sonochem. 16, 488–94.

Lenz, M., Hullebusch, E.D. Van, Hommes, G., Corvini, P.F.X., Lens, P.N.L., 2008. Selenate removal in methanogenic and sulfate-reducing upflow anaerobic sludge bed reactors. Water Res. 42, 2184–2194.

Lenz, M., Lens, P.N.L., 2009. The essential toxin: the changing perception of selenium in environmental sciences. Sci. Total Environ. 407, 3620–3633.

Li, X.Y., Yang, S.F., 2007. Influence of loosely bound extracellular polymeric substances (EPS) on the flocculation, sedimentation and dewaterability of activated sludge. Water Res. 41, 1022–1030.

Martinez, R.E., Smith, D.S., Kulczycki, E., Ferris, F.G., 2002. Determination of intrinsic bacterial surface acidity constants using a donnan shell model and a continuous pK(a) distribution method. J. Colloid Interface Sci. 253, 130–139.

McKay, G., Porter, J.F., 1997. Equilibrium Parameters for the Sorption of Copper, Cadmium and Zinc Ions onto Peat. J. Chem. Technol. Biotechnol. 69, 309–320.

Nakamura, K., Ikawa, A., 2001. Infrared absorption in amorphous selenium. Comput. Phys. Commun. 142, 295–299.

Oremland, R.S., Herbel, M.J., Blum, J.S., Langley, S., Beveridge, T.J., Ajayan, P.M., Sutto, T., Ellis, A. V, Curran, S., 2004. Structural and spectral features of selenium nanospheres produced by Se-respiring bacteria. Appl. Environ. Microbiol. 70, 52–60.

Pearce, C.I., Pattrick, R. a D., Law, N., Charnock, J.M., Coker, V.S., Fellowes, J.W., Oremland, R.S., Lloyd, J.R., 2009. Investigating different mechanisms for biogenic selenite transformations: *Geobacter sulfurreducens, Shewanella oneidensis and Veillonella atypica*. Environ. Technol. 30, 1313–1326.

Rayman, M.P., 2000. The importance of selenium to human health. Lancet 356, 233–241.

Turner, B.F., Fein, J.B., 2006. Protofit: A program for determining surface protonation constants from titration data. Comput. Geosci. 32, 1344–1356.

Wang, L.-L., Wang, L.-F., Ren, X.-M., Ye, X.-D., Li, W.-W., Yuan, S.-J., Sun, M., Sheng, G.-P., Yu, H.-Q., Wang, X.-K., 2012. pH dependence of structure and surface properties of microbial EPS. Environ. Sci. Technol. 46, 737–744.

Wang, X., Liang, X., Wang, Y., Wang, X., Liu, M., Yin, D., Xia, S., Zhao, J., Zhang, Y., 2011. Adsorption of Copper (II) onto activated carbons from sewage sludge by microwave-induced phosphoric acid and zinc chloride activation. Desalination 278, 231–237.

Winkel, L.H.E., Johnson, C.A., Lenz, M., Grundl, T., Leupin, O.X., Amini, M., Charlet, L., 2012. Environmental selenium research: from microscopic processes to global understanding. Environ. Sci. Technol. 46, 571–579.

Wright, M.T., Parker, D.R., Amrhein, C., 2003. Critical evaluation of the ability of sequential extraction procedures to quantify discrete forms of selenium in sediments and soils. Environ. Sci. Technol. 37, 4709–4716.

Wu, L., 2004. Review of 15 years of research on ecotoxicology and remediation of land contaminated by agricultural drainage sediment rich in selenium. Ecotoxicol. Environ. Saf. 57, 257–269.

Xu, C., Zhang, S., Chuang, C., Miller, E.J., Schwehr, K. a., Santschi, P.H., 2011. Chemical composition and relative hydrophobicity of microbial exopolymeric substances (EPS) isolated by anion exchange chromatography and their actinide-binding affinities. Mar. Chem. 126, 27–36.

Xu, C.L., Wang, Y.Z., Jin, M.L., Yang, X.Q., 2009. Preparation, characterization and immunomodulatory activity of selenium-enriched exopolysaccharide produced by bacterium *Enterobacter cloacae* Z0206. Bioresour. Technol. 100, 2095–7. doi:10.1016/j.biortech.2008.10.037

Ye, F., Ye, Y., Li, Y., 2011. Effect of C/N ratio on extracellular polymeric substances (EPS) and physicochemical properties of activated sludge flocs. J. Hazard. Mater. 188, 37–43.

Zhang, W., Chen, Z., Liu, H., Zhang, L., Gao, P., Li, D., 2011. Biosynthesis and structural characteristics of selenium nanoparticles by *Pseudomonas alcaliphila*. Colloids Surf. B. Biointerfaces 88, 196–201.

CHAPTER 11

Discussion

Insights into the properties of BioSeNPs to better manage selenium

Role of EPS

❑ Present on the surface
❑ Provide colloidal stability
❑ Interface to heavy metals and provides selectivity towards heavy metals
❑ Improves engineering properties of activated sludge

Role of thermophilic temperatures

❑ Transform to trigonal nanowire crystallinity and morphology
❑ Better retention of elemental selenium in an UASB reactor

11.1 Summary of the results

The main objective of this thesis was to study biogenic elemental selenium nanoparticles (BioSeNPs) produced by anaerobic granular sludge and activated sludge systems. The three subobjectives were: BioSeNPs production and characterization, BioSeNPs' application in heavy metal removal and fate of BioSeNPs in the bioreactors. All three sub objectives have provided valuable insight into the properties of BioSeNPs that can be exploited to better treat selenium rich wastewaters and develop recovery systems.

In chapter 3, the extracellular polymeric substances (EPS) were the organic layer present on the surface of BioSeNPs. EPS was found to govern the surface charge of the BioSeNPs. It was also observed that EPS can effectively cap the elemental selenium nanoparticles during their chemical synthesis (CheSeNPs). The EPS capped CheSeNPs were spherical nanospheres as compared to nanorods formed in the absence of EPS at ambient temperature. Biogenic elemental selenium nanowires (BioSeNWs) were produced at more elevated temperatures (55-60 ^{0}C), as demonstrated by incubations performed at these temperatures in chapter 4. The produced BioSeNWs were of similar shape and size as synthesized by chemical production methods. The BioSeNWs were colloidally stable and were capped by EPS as opposed to CheSeNPs produced in absence of EPS. The ζ-potential versus pH profile of BioSeNWs was similar to the one observed for EPS, EPS capped CheSeNPs and BioSeNPs suggesting that EPS govern the surface charge of BioSeNWs as well.

In chapter 6, adsorption of Zn, as a model divalent heavy metal, onto BioSeNPs was investigated. The adsorption of Zn onto BioSeNPs at acidic pH values follow a ligand-like (type II) adsorption mechanism. The adsorption of Zn onto BioSeNPs followed a two-step process at near-neutral pH. X-ray photoelectron spectroscopy (XPS) suggested the precipitation of one of the Zn species which could have been ZnO, ZnSe, $Zn(OH)_2$, or $ZnCO_3$. Preliminary Extended X-ray Absorption of Fine Structure (EXAFS) data analysis suggests the absence of ZnSe, thus discarding the occurrence of disproportionation of elemental selenium. The ζ-potential of BioSeNPs became lesser negative at higher loading of Zn, resulting in lower colloidal stability, thus leading to higher retention of BioSeNPs on the filter when compared to BioSeNPs after filtration but without adsorption.

The selective adsorption of heavy metals onto BioSeNPs was explored in chapter 7. It was found that the metals to BioSeNPs ratio (v:v) and pH can be manipulated to optimize the selective adsorption of Cu. At the metal to BioSeNPs ratio of 1:1 (v:v) and theoretical pH of 4.3, Cu was found to adsorb 4.7 times more onto BioSeNPs than the total sum of Cd and Zn adsorbed, when an equimolar mixture of Cu, Cd and Zn was used in the adsorption experiments. The selective preference of Cu onto BioSeNPs depends on the intrinsic properties of Cu: smaller ionic radius, higher electronegativity, higher ratio of ionization potential to ionic radius, higher first stability constant of metal hydroxo species and acetate complexes. The selective preference of Cu also depends on the presence of functional groups such as hydroxyl and carboxyl on the surface of BioSeNPs. Indeed, FT-IR analysis of BioSeNPs loaded with heavy metals confirmed the interaction of hydroxl and carboxyl groups on the surface of BioSeNPs with heavy metals.

Chapter 8 explored the use of thermophilic conditions to retain the produced elemental selenium after the reduction of selenate in an upflow anaerobic sludge blanket reactor (UASB). The total selenium concentration in the effluent of thermophilic reactors was lower than the one measured in the mesophilic reactor when the feed selenate concentration was 50 µM. However, the dissolved selenium concentration in the effluent of both the reactors was similar. This suggested that the produced elemental selenium was better retained in the thermophilic reactor as compared to the mesophilic reactor. When nitrate was fed in the influent, dissolved selenium concentrations remained the same but there was an increase in the total selenium concentration in the mesophilic reactor, suggesting extracellular production of elemental selenium. The presence of nitrate in the feed deteriorated the performance of thermophilic UASB reactor. However, the removal efficiency of total selenium and dissolved selenium increased by 6% when the nitrate feed was further increased, suggesting that thermophilic reactors can also be adapted for higher selenium removal efficiencies at high nitrate to selenate ratios.

Aerobic removal of selenite by activated sludge, as explained in Chapter 9, can be exploited to develop a single-step process for selenium removal. The higher airflow rate and lower temperature displayed better entrapment of selenium in the biomass. This was due to the fact the intracellular production of elemental selenium is enhanced under aerobic conditions, in contrast to selenite reduction under micro-aerobic or anaerobic that mainly takes place in the periplasmic space or extracellularly. It was also observed that glucose was by far the best substrate to reduce selenite under aerobic conditions. The selenium fed activated sludge showed better settleability and hydrophilicity, but poorer dewaterability at higher total suspended solids (TSS) concentrations as compared to the control activated sludge. The selenium fed activated sludge also showed a less negative surface charge density as compared to the control activated sludge. This all can be ascribed to presence of trapped elemental selenium in the biomass, as shown by sequential extraction procedures, scanning/ transmission electron microscopy and X-ray diffraction pattern. The other possible reason can be the higher amount and different composition of EPS generated by selenium fed activated sludge as compared to the control activated sludge.

11.2 Role of EPS in the fate of selenium in environment and bioreactors

EPS is mainly composed of polysaccharides, proteins, humic substances, lipids and nucleic acids (D'Abzac et al., 2010; Sheng et al., 2010). Proteins and polysaccharides are the major components of EPS. Generally, EPS is known to retard or prevent the dispersion of nanomaterials such as silver nanoparticles (Kang et al., 2014; Tourney and Ngwenya, 2014). In contrast, this study (Chapter 3) showed that the EPS provides colloidal stability to the BioSeNPs due to less negative ζ-potential values (Buchs et al., 2013; Dhanjal and Cameotra, 2010; Jain et al., 2015). Bare elemental selenium has been reported to have a ζ-potential of -10 mV as compared to -30 mV observed for BioSeNPs (Dhanjal and Cameotra, 2010; Feng et al., 2013; Jain et al., 2015). This colloidal stability of the BioSeNPs is the reason for their presence in bioreactors effluents as well as the high mobility of BioSeNPs in the environment. The role of the EPS in the formation and characteristics of BioSeNPs has been demonstrated in Figure 11.1.

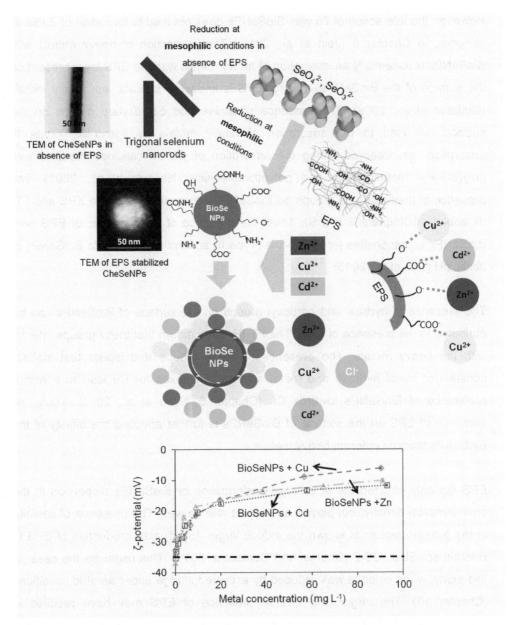

Figure 11.1. Role of EPS in determining the properties of the BioSeNPs.

The presence of EPS on the surface of BioSeNPs determines the mechanism of interaction of BioSeNPs with heavy metals. The capture of mercury from the vapor phase by elemental selenium is due to the precipitation of mercury selenide on the surface of the elemental selenium (Fellowes et al., 2011; Johnson et al., 2008).

However, the interaction of Zn with BioSeNPs does not lead to formation of ZnSe as observed in Chapter 6 (Jain et al., 2015). The interaction of heavy metals with BioSeNPs is essentially an interaction of heavy metals with the EPS layer present on the surface of the BioSeNPs. Indeed, EPS is known to interact with heavy metals (Guibaud et al., 2009). The presence of amine and carboxylate groups on the surface can lead to the adsorption of heavy metals by ligand-like (Type II) adsorption, as observed during the adsorption of Cu on cellulose modified with poly(glycidyl methacrylate) and polyethyleneimine (Navarro et al., 2001). The presence of these surface groups on BioSeNPs was confirmed by the XPS and FT-IR analysis (Chapters 3 and 6). Thus, the presence of such a layer of EPS onto BioSeNPs is responsible for ligand-like (Type II) adsorption of Zn onto BioSeNPs at acidic pH (Jain et al., 2015).

The presence of hydroxl and carboxyl groups on the surface of BioSeNPs can be attributed to the presence of EPS. The FT-IR data confirm that these groups interact with the heavy metals. The presence of these groups and higher first stability constant of metal hydroxo and metal acetate complexes for Cu lead to a higher preference of BioSeNPs towards Cu (Chapter 6, Sitko et al., 2013). Thus, the presence of EPS on the surface of BioSeNPs is further affecting the affinity of the BioSeNPs towards different heavy metals.

EPS not only effected the selenium transformation or BioSeNPs dispersion in the environmental directly, but played an indirect role as well. The presence of selenite in the growth medium is known the induce larger and diverse production of EPS in *Bacillus* sp. Strain JS-2 (Dhanjal and Cameotra, 2011). This might be the case, in this study, when selenite was reduced by activated sludge under aerobic conditions (Chapter 10). The larger and diverse presence of EPS may have resulted in improved hydrophilicity of selenium fed activated sludge as compared to control activated sludge (Chapter 10). Indeed, the variation of EPS has an effect on the hydrophilicity of the activated sludge (Sheng et al., 2010; Tourney and Ngwenya, 2014). Similarly, a lesser negative surface charge of selenium fed activated sludge as compared to the control activated sludge can be attributed to larger number of carboxyl and hydroxyl groups in the EPS of selenium fed activated sludge (Tourney and Ngwenya, 2014, 2010).

11.3 Role of temperature in determining selenium properties and its transformation

The most direct effect of temperature on the properties of biogenic elemental was the transformation of amorphous or monoclinic red colored elemental selenium to trigonal gray elemental selenium nanowires as observed in chapter 4. The use of thermophilic conditions (55 $^\circ$C and 65 $^\circ$C) lead to the formation BioSeNWs whose shape and size seems similar to the ones observed in study by Gates et al. (2002). This transformation to BioSeNWs was also observed during the thermophilic reduction of selenate in an UASB reactor (Chapter 8). Though the surface charge on BioSeNPs and BioSeNWs is similar, the difference in the shape and size of these BioSeNWs would certainly lead to differences in their capacity to adsorb metals ions. This might be the reason of a higher retention of elemental selenium in the thermophilic UASB reactor (Chapter 8). The surface charge on BioSeNWs is due to the presence of EPS on the surface. It is interesting to note that when CheSeNPs have been formed at room temperature in the absence of EPS, the produced CheSeNPs transformed to the trigonal phase. The transformation to trigonal crystalline state is thermodynamic favorable, however, it does not take place at 30 $^\circ$C for BioSeNPs (Oremland et al., 2004; Wang et al., 2010). However, for the case of BioSeNPs, this transformation takes place at much more elevated temperature (55 and 65 $^\circ$C, Chapter 4). This might be due to the presence of EPS that inhibits or retard this transformation.

Elevated temperature also has an indirect effect on the transformation of selenate to elemental selenium. The anaerobic granular sludge showed higher adaptation time to reduce selenate at elevated temperature and low influent selenate concentration (Chapter 8). Another interesting effect of elevated temperature was the non-release of elemental selenium in the effluent as observed when nitrate was introduced at mesophilic temperatures. This might be due to a shift in the micro-organisms that produce BioSeNPs and thus, differences in selenate reduction mechanisms (Chapter 8). The effect of temperature on the morphology of elemental selenium nanoparticles and their fate in the bioreactors is summarized in Figure 11.2.

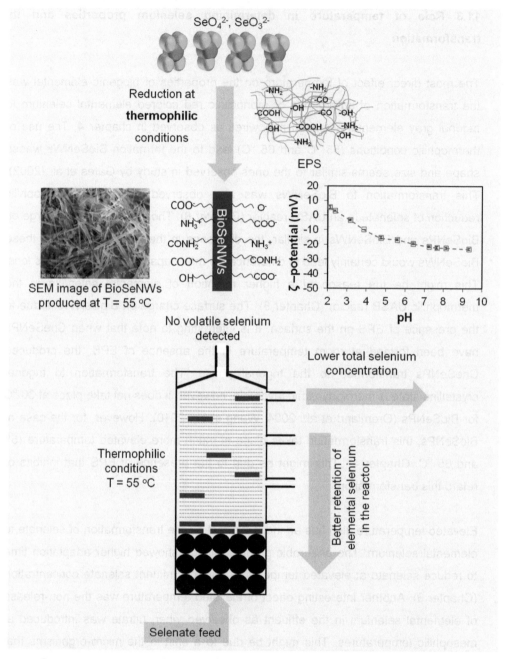

Figure 11.2 Effect of thermophilic conditions on the transformation of selenium.

11.4 Future perspectives

The microbial interactions with selenium and its transformation to elemental selenium were studied in this thesis. This study improved our understanding on the formation, properties and fate of the BioSeNPs in the bioreactors. This study also shed light on the effect of temperature on the reduction of selenate and transformation to trigonal elemental selenium.

There are still many areas identified by this study that needs to be explored. One of line of research that needs to be explored is the manipulation of EPS to change the surface charge of the BioSeNPs. The change in the surface charge of BioSeNPs can be used to selectively adsorb cations and also possibly anions. This might be achieved by shifting of iso-electric point. The protein to polysaccharide ratio in the EPS needs to be manipulated to change the surface charge of the EPS and consequently the iso-electric point of BioSeNPs (More et al., 2014). The manipulation of the protein to polysaccharide can be manipulated by means of changing the C/N ratio in the feed (Ye et al., 2011).

Another phenomenon that still not fully understood is the intracellular or extracellular production of BioSeNPs. This understanding would help to improve the treatment processes and better manage selenium recovery. For example, a complete intracellular BioSeNPs production, as observed in the case of activated sludge at higher dissolved oxygen levels in Chapter 9, would make the additional step required for the treatment of the effluent of an UASB reactor redundant (Buchs et al., 2013; Jain et al., 2015; Staicu et al., 2014). The activated sludge that is trapping BioSeNPs can then be used to recover selenium or used as a slow selenium fertilizer for the fortification of the crops (Bhatia et al., 2013; Haug et al., 2007). On the other hand, for the application of BioSeNPs in material science, it will be of interest that reduction of selenium oxyanions takes place extracellularly. This was observed for the reduction of selenite by *Shewanella oneidensis* MR-1(Li et al., 2014). However, the exact reduction mechanisms and the ability to manipulate those mechanisms to trigger the extracellular or intracellular production of BioSeNPs are not well understood.

The chemical reactivity of BioSeNPs with other elements, especially heavy metals, is not well known. For example, elemental selenium can disproportionate to form selenide and selenite upon interaction with heavy metals. The likelihood of disproportionation is based on the metal selenide solubility product (Li et al., 1999; Nuttall, 1987). The disproportionation of BioSeNPs is important as the products of this reaction: metal selenide and selenite, both are more mobile and toxic than the BioSeNPs (Winkel et al., 2012). However, there is no research so far to explore the disproportionation of selenium with heavy metals other than Zn (Chapter 6).

11.5 References

Bhatia, P., Aureli, F., D'Amato, M., Prakash, R., Cameotra, S.S., Nagaraja, T.P., Cubadda, F., 2013. Selenium bioaccessibility and speciation in biofortified Pleurotus mushrooms grown on selenium-rich agricultural residues. Food Chem. 140, 225–230.

Buchs, B., Evangelou, M.W.-H., Winkel, L., Lenz, M., 2013. Colloidal properties of nanoparticular biogenic selenium govern environmental fate and bioremediation effectiveness. Environ. Sci. Technol. 47, 2401–2407.

D'Abzac, P., Bordas, F., Van Hullebusch, E., Lens, P.N.L., Guibaud, G., 2010. Extraction of extracellular polymeric substances (EPS) from anaerobic granular sludges: comparison of chemical and physical extraction protocols. Appl. Microbiol. Biotechnol. 85, 1589–1599.

Dhanjal, S., Cameotra, S.S., 2010. Aerobic biogenesis of selenium nanospheres by Bacillus cereus isolated from coalmine soil. Microb. Cell Fact. 9, 52.

Dhanjal, S., Cameotra, S.S., 2011. Selenite Stress Elicits Physiological Adaptations in Bacillus sp . (Strain JS-2). J. Microbiol. Biotechnol. 21, 1184–1192.

Fellowes, J.W., Pattrick, R.A.D., Green, D.I., Dent, A., Lloyd, J.R., Pearce, C.I., 2011. Use of biogenic and abiotic elemental selenium nanospheres to sequester elemental mercury released from mercury contaminated museum specimens. J. Hazard. Mater. 189, 660–669.

Feng, Y., Su, J., Zhao, Z., Zheng, W., Wu, H., Zhang, Y., Chen, T., 2013. Differential effects of amino acid surface decoration on the anticancer efficacy of selenium nanoparticles. Dalton Trans. 43, 1854–1861.

Gates, B., Mayers, B., Cattle, B., Xia, Y., 2002. Synthesis and characterization of uniform nanowires of trigonal Selenium. Adv. Funct. Mater. 12, 219–227.

Guibaud, G., van Hullebusch, E., Bordas, F., d'Abzac, P., Joussein, E., 2009. Sorption of Cd(II) and Pb(II) by exopolymeric substances (EPS) extracted from activated sludges and pure bacterial strains: modeling of the metal/ligand ratio effect and role of the mineral fraction. Bioresour. Technol. 100, 2959–2968.

Haug, A., Graham, R.D., Christophersen, O.A., Lyons, G.H., 2007. How to use the world's scarce selenium resources efficiently to increase the selenium concentration in food. Microb. Ecol. Health Dis. 19, 209–228.

Jain, R., Jordan, N., Schild, D., Hullebusch, E.D. Van, Weiss, S., Franzen, C., Hubner, R., Farges, F., Lens, P.N.L., 2015. Adsorption of zinc by biogenic elemental selenium nanoparticles. Chem. Eng. J. 260, 850–863.

Johnson, N.C., Manchester, S., Sarin, L., Gao, Y., Kulaots, I., Hurt, R.H., 2008. Mercury vapor release from broken compact fluorescent lamps and in situ capture by new nanomaterial sorbents. Environ. Sci. Technol. 42, 5772–5778.

Kang, F., Alvarez, P.J., Zhu, D., 2014. Microbial extracellular polymeric substances reduce Ag+ to silver nanoparticles and antagonize bactericidal activity. Environ. Sci. Technol. 48, 316–322.

Li, D.-B., Cheng, Y.-Y., Wu, C., Li, W.-W., Li, N., Yang, Z.-C., Tong, Z.-H., Yu, H.-Q., 2014. Selenite reduction by Shewanella oneidensis MR-1 is mediated by fumarate reductase in periplasm. Sci. Rep. 4, 3735.

Li, Y., Ding, Y., Liao, H., Qian, Y., 1999. Room-temperature conversion route to nanocrystalline mercury chalcogenides HgE (E=S,Se,Te). J. Phys. Chem. Solids 60, 965–968.

More, T.T., Yadav, J.S.S., Yan, S., Tyagi, R.D., Surampalli, R.Y., 2014. Extracellular polymeric substances of bacteria and their potential environmental applications. J. Environ. Manage. 144, 1–25.

Navarro, R.R., Tatsumi, K., Sumi, K., Matsumura, M., 2001. Role of anions on heavy metal sorption of a cellulose modified with poly(glycidyl methacrylate) and polyethyleneimine. Water Res. 35, 2724–30.

Nuttall, K.L., 1987. A model for metal selenide formation under biological conditions. Medical Hyptheses 24, 217–221.

Oremland, R.S., Herbel, M.J., Blum, J.S., Langley, S., Beveridge, T.J., Ajayan, P.M., Sutto, T., Ellis, A. V, Curran, S., 2004. Structural and spectral features of selenium nanospheres produced by Se-respiring bacteria. Appl. Environ. Microbiol. 70, 52–60.

Sheng, G.-P., Yu, H.-Q., Li, X.-Y., 2010. Extracellular polymeric substances (EPS) of microbial aggregates in biological wastewater treatment systems: a review. Biotechnol. Adv. 28, 882–894.

Sitko, R., Turek, E., Zawisza, B., Malicka, E., Talik, E., Heimann, J., Gagor, A., Feist, B., Wrzalik, R., 2013. Adsorption of divalent metal ions from aqueous solutions using graphene oxide. Dalton Trans. 42, 5682–5689.

Staicu, L.C., van Hullebusch, E.D., Lens, P.N.L., Pilon-Smits, E.A., Oturan, M. a, 2014. Electrocoagulation of colloidal biogenic selenium. Environ. Sci. Pollut. Res. Int. (accepted). doi:10.1007/s11356-014-3592-2

Tourney, J., Ngwenya, B.T., 2010. The effect of ionic strength on the electrophoretic mobility and protonation constants of an EPS-producing bacterial strain. J. Colloid Interface Sci. 348, 348–354.

Tourney, J., Ngwenya, B.T., 2014. The role of bacterial extracellular polymeric substances in geomicrobiology. Chem. Geol. 386, 115–132.

Wang, T., Yang, L., Zhang, B., Liu, J., 2010. Extracellular biosynthesis and transformation of selenium nanoparticles and application in H_2O_2 biosensor. Colloids Surf. B. Biointerfaces 80, 94–102.

Winkel, L.H.E., Johnson, C.A., Lenz, M., Grundl, T., Leupin, O.X., Amini, M., Charlet, L., 2012. Environmental selenium research: from microscopic processes to global understanding. Environ. Sci. Technol. 46, 571–579.

Ye, F., Ye, Y., Li, Y., 2011. Effect of C/N ratio on extracellular polymeric substances (EPS) and physicochemical properties of activated sludge flocs. J. Hazard. Mater. 188, 37–43.

Appendix 1

This appendix is submitted as supporting information for:

Jain, R., Jordan, N., Weiss, S., Foerstendorf, H., Heim, K., Kacker, R., Hübner, R., Kramer, H., Hullebusch, E.D. Van, Farges, F., Lens, P.N.L., 2015. Extracellular polymeric substances (EPS) govern the surface charge of biogenic elemental selenium nanoparticles (BioSeNPs). Environ. Sci. Tech. 49, 1713-1720

ζ-potential measurements for EPS and BSA capped CheSeNPs loaded with Zn

EPS and BSA-capped CheSeNPs (50 mg L^{-1}) were produced and purified by dialysis as described in the manuscript. The pH of the EPS and BSA-capped CheSeNPs was changed to 7.3 using 1 M NaOH. 0.5 mL of $ZnCl_2$ was added to 5.0 mL of EPS and BSA-capped CheSeNPs to vary the Zn concentration from 50 to 1000 mg L^{-1}. The final pH of the EPS and BSA-capped CheSeNPs varied from 5.5 to 6.5. The ζ-potential of EPS and BSA-capped CheSeNPs loaded with Zn were measured in triplicates.

Analytics

SEM-EDXS

To characterize the surface morphology of the BioSeNPs, scanning electron microscopy (SEM) was performed using a S-4800 microscope (Hitachi) operated at an accelerating voltage of 10 kV. For qualitative chemical analysis of the BioSeNPs, energy-dispersive X-ray spectroscopy (EDXS) analysis was carried out by means of a conventional Si(Li) detector with S-UTW window (Oxford Instruments) attached to the SEM. Sample preparation was done by spreading a small amount of BioSeNPs solution over a piece of a silicon wafer, drying it for a few hours at room temperature and mounting the sample on an aluminum holder for SEM analysis.

ζ-potential and hydrodynamic diameter measurements

The ζ-potential and hydrodynamic diameter (HDD) were calculated by DTS software (Malvern Instrument) using electrophoretic mobility and dynamic light scattering measurements carried out at 22 °C by a Nano Zetasizer (Malvern Instruments) at a laser beam of 633 nm and a scattering angle of 173°. The refractive index of 2.6 for selenium was used in the HDD measurement (Dobias et al., 2011). As the concentration of selenium nanoparticles in water was low, viscosity of water at 22 °C was used for the measurements. The general purpose algorithm in the DTS software was used for calculating the size distribution.

FT-IR spectroscopy

For IR spectroscopy, KBr pellets were prepared by mixing approximately 1 mg of the samples with 300 mg dried KBr and subsequent pressing for 2 minutes at 145,000 psi until clear pellets were obtained. The FT-IR spectra of BioSeNPs and CheSeNPs were carried out on a Bruker Vertex 70/v spectrometer equipped with a D-LaTGS-detector (L-alanine doped triglycine sulfate), over the range 4000-400 cm^{-1} in the transmittance mode, with a spectral resolution of 4 cm^{-1}. Each spectrum was averaged out over 64 scans.

Acid-base titration

To determine the pKs of BioSeNPs, acid-base titration was carried out using a Metrohm autotitrator unit. 0.1966 mg of BioSeNPs was used in a total volume of 30 mL with a background electrolyte concentration of 1 mM NaCl. The initial pH raised above 9.4 by addition of 0.102 M NaOH. The BioSeNPs were continuously stirred and flushed with nitrogen. The titration was carried by automatic addition of 0.1 mL of HCl (0.01214 M). The change in background ionic strength due to the addition of acid was less than 8%. For the control titration, Milli-Q water (18MΩ cm) at 1 mM of background electrolyte concentration was used.

Total organic carbon analyzer

The extracted EPS was characterized by total organic carbon and total nitrogen measurements using a Shimadzu TOC-VCPN analyzer. Prior to analysis, the samples were filtered with 0.45 μm filters (Whatman, Dassel, Germany). The determined dissolved organic carbon was considered as the total organic carbon.

Fluorescence excitation and emission matrix spectroscopy

EPS was characterized for various components using a FluoroMax-3 spectrofluorometer (HORIBA Jobin Yvon, Edison, NJ, USA). The samples were diluted to bring the dissolved organic carbon concentration below 1 mg L^{-1}. The fluorometer was operated and stabilized as described in a previous study (Maeng et al., 2012). The measurements were carried out at excitation and emission wavelengths of 200-400 nm and 300-500 nm, respectively.

TEM-EDXS

Transmission electron microscopy investigations were performed to locally analyze the microstructure and in particular the morphology of the EPS and BSA-capped CheSeNPs and CheSeNPs formed in the absence of EPS or BSA. An image-corrected Titan 80-300 microscope (FEI) operated at an accelerating voltage of 300 kV was used. For sample preparation, one droplet of nanoparticles suspended in water was deposited onto a 400 mesh Cu grid coated with a carbon support film. After drying in a desiccator at room temperature and covering with an additional carbon-coated Cu grid, the TEM specimen was placed into a double-tilt analytical holder to perform the TEM analyses.

Figures and tables

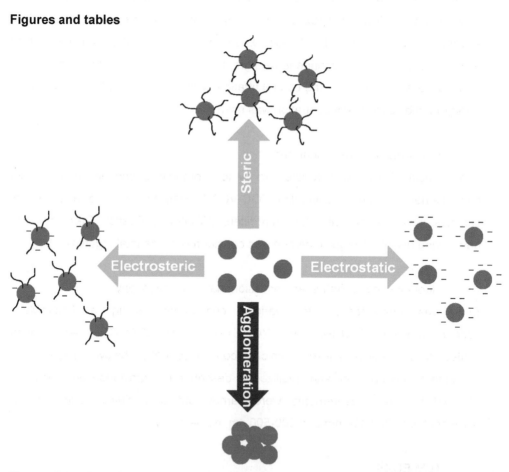

Figure S1. Graphical representation of various stabilization mechanisms of the nanoparticles inhibiting agglomeration.

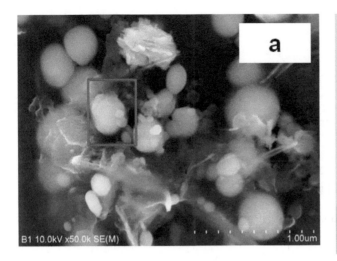

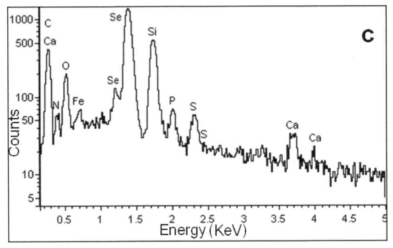

Figure S2. Secondary electron SEM images of (a) BioSeNPs, (b) colloidal suspension of BioSeNPs and (c) representative EDX spectra confirming the presence of selenium in BioSeNPs from the area marked by red square in (a). Please note that the samples were deposited onto a piece of Si wafer.

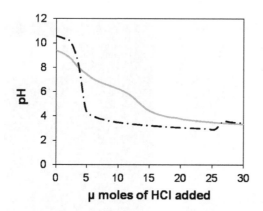

Figure S3. Acid-base titration of BioSeNPs produced at 30 °C (—) and MQ water as control (—·—·—).

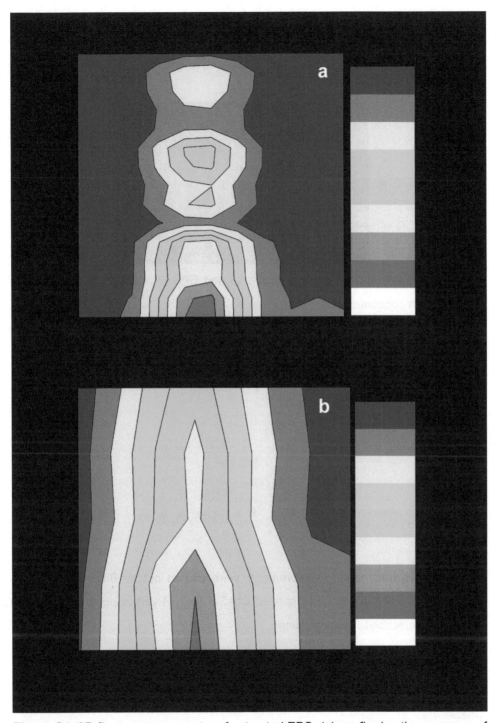

Figure S4. 3D fluorescence spectra of extracted EPS: (a) confirming the presence of aromatic proteins and soluble microbial byproduct by observing a maxima at excitation and emission wavelength of 230/370 nm and excitation and emission wavelength of 300/370 nm, respectively; and (b) further aromatic proteins by observing a maxima at excitation and emission wavelength of 230/330 nm.

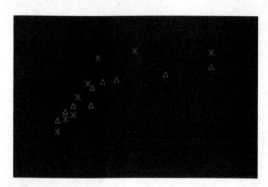

Figure S5. ζ-potential variation of BSA (*) and EPS (Δ) capped CheSeNPs with increasing Zn concentrations.

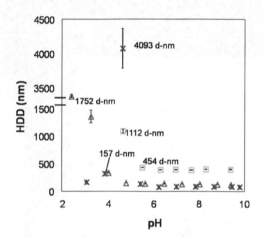

Figure S6. Hydrodynamic measurements were carried out for BioSeNP (□), EPS capped CheSeNPs (Δ) and BSA capped CheSeNPs (*) versus pH at 10 mM NaCl background electrolyte concentrations.

Table S1. Assignments of various functional groups to different features (cm^{-1}) in the FT-IR spectra of BioSeNPs, EPS, EPS capped CheSeNPs, BSA and BSA capped CheSeNPs.

Functional groups	BioSeNPs	EPS	EPS capped CheSeNPs	BSA	BSA capped CheSeNPs	Ref.
-O-H, -N-H	3404-3270	3400, 3062	3420, 3062	3315, 3063	3283, 3063	(Xu et al., 2011)
-C-H	2959, 2928, 2866	2962, 2932	2928	2959, 2932, 2866	2955, 2934, 2865	(Wang et al., 2012)
-COOH	1720	1720	1720	–	1720	
-C=O Amide-I	1646	1653	1646	1654	1654	(Xu et al., 2011)
-N-H Amide-II	1542	1530	1537	1532	1540	(Wang et al., 2012)
-CH$_3$/-COO$^-$ antisymmetric	1460	1452	-	1451	-	(Zhu et al., 2012)
-COO$^-$ symmetric	1394	1388	1404	1390	1401	(Zhu et al., 2012)
-C-N, -N-H, P=O	1242	1236	1243	1242	1238	(Wang et al., 2012)
-P–O	1151	1153	1151	1166	-	(Wang et al., 2012)
-C-O-C, -C-H	1073 1038	-	1077-1040	1077-1040	-	-	(Wang et al., 2012), (Zhu et al., 2012)

References

Dobias, J., Suvorova, E.I., Bernier-latmani, R., 2011. Role of proteins in controlling selenium nanoparticle size. Nanotechnology 22, 195605.

Maeng, S.K., Sharma, S.K., Abel, C.D.T., Magic-Knezev, A., Song, K.-G., Amy, G.L., 2012. Effects of effluent organic matter characteristics on the removal of bulk organic matter and selected pharmaceutically active compounds during managed aquifer recharge: Column study. J. Contam. Hydrol. 140-141, 139–49.

Wang, L.-L., Wang, L.-F., Ren, X.-M., Ye, X.-D., Li, W.-W., Yuan, S.-J., Sun, M., Sheng, G.-P., Yu, H.-Q., Wang, X.-K., 2012. pH dependence of structure and surface properties of microbial EPS. Environ. Sci. Technol. 46, 737–44.

Xu, C., Zhang, S., Chuang, C., Miller, E.J., Schwehr, K. a., Santschi, P.H., 2011. Chemical composition and relative hydrophobicity of microbial exopolymeric substances (EPS) isolated by anion exchange chromatography and their actinide-binding affinities. Mar. Chem. 126, 27–36.

Zhu, L., Qi, H., Lv, M., Kong, Y., Yu, Y., Xu, X., 2012. Component analysis of extracellular polymeric substances (EPS) during aerobic sludge granulation using FTIR and 3D-EEM technologies. Bioresour. Technol. 124, 455–9.

Appendix 2

This appendix is submitted as supporting information for:

Jain, R., Jordan, N., Schild, D., Hullebusch, E.D. Van, Weiss, S., Franzen, C., Hubner, R., Farges, F., Lens, P.N.L., 2015. Adsorption of zinc by biogenic elemental selenium nanoparticles. Chem. Eng. J. 260, 850–863.

BioSeNPs concentration determination

The biologically produced elemental selenium nanoparticles (BioSeNPs) concentration was measured by ICP-MS by dissolving the sample in concentrated HNO_3. The measurements were carried out with the reaction gas H_2:He (7:93) and used [78]Se for quantification and [80]Se for verification. The samples were prepared in 0.5% HNO_3 and injected in a 1:1 ratio with internal standards of Li, Ga, Sc, Rh and Ir. The samples were measured in triplicate and the entire system was flushed with ultrapure 0.5% HNO_3 in MilliQ (18MΩ*cm) water.

Metal determination

The residual zinc ions were measured by Atomic Absorption Spectroscopy (AAS200, PerkinElmer) at 213.9 nm. Calcium, magnesium and iron were measured using AAS200 at 422.8, 285.2 and 248.3 nm, respectively.

Chloride ion measurements

Anion (Cl^-, NO_3^-, PO_4^{3-} and SO_4^{2-}) concentrations were measured by an ion chromatograph (Dionex ICS 1000) using an IonPac As-14A anion-exchange column using carbonate/bicarbonate eluent coupled with suppressed conductivity detection at a flow rate of 0.5 mL min^{-1}.

BioSeNPs characterization

SEM-EDXS

To characterize the surface morphology of the BioSeNPs, scanning electron microscopy (SEM) was performed using a S-4800 microscope (Hitachi) operated at an accelerating voltage of 10 kV. For qualitative chemical analysis of the BioSeNPs, energy-dispersive X-ray spectroscopy (EDXS) analysis was carried out by means of a conventional Si(Li) detector with super ultrathin window (Oxford Instruments) attached to the SEM. Sample preparation was done by spreading a small amount of the BioSeNPs solution over a piece of a silicon wafer, drying it for a few hours at room temperature and mounting the sample on an aluminum holder for SEM analysis.

Electrophoretic mobility measurements

The ζ-potential was calculated by DTS software (Malvern Instrument) using electrophoretic measurements carried out at 23 °C by a Nano Zetasizer (Malvern instrument) at a laser beam of 633 nm and a scattering angle of 173°.

X-ray diffraction

X-ray diffraction (XRD) analysis was performed on a Bruker D8 Advance diffractometer equipped with an energy dispersion Sol-X detector with copper radiation (CuKα, λ = 0.15406 nm). The acquisition was recorded between 2° and 80°, with a 0.02° scan step and 1 s step time. Samples were spread over the sample holder and dried at room temperature.

XPS measurements

The uppermost surface layers (up to ~10 nm) of BioSeNPs as well as zinc ion contacted BioSeNPs were analyzed by X-ray Photoelectron Spectroscopy (XPS) at room temperature. XPS analysis was carried out by a XP spectrometer (ULVAC-PHI, Inc., model PHI 5000 VersaProbe II). A scanning microprobe X-ray source (monochromatic Al Kα (1486.6 eV) X-rays) was applied in combination with low energy electrons and low energy Ar ions for charge compensation (dual beam technique). The spectrometer is equipped with a hemispherical capacitor analyzer (mean diameter 279.4 mm), and the detector consists of a microchannel detector with 16 anodes. Calibration of the binding energy scale of the spectrometer was performed using well-established binding energies of elemental lines of pure metals (monochromatic Al Kα: Cu $2p_{3/2}$ at 932.62 eV, Au $4f_{7/2}$ at 83.96 eV) (Seah et al., 1998). Standard deviations of binding energies of isolating samples were within ±0.2 eV.

Samples were centrifuged and decanted. Aliquots were deposited onto an indium foil and mounted on a stainless steel sample holder under anoxic conditions (glovebox with Ar atmosphere). Samples were transferred into the analysis chamber of the XPS using a vacuum transfer vessel (without air contact).

Spectra were collected at a take-off angle of 45° (angle between sample surface and analyzer) and the pressure inside the spectrometer was about 2×10^{-7} Pa. To retrieve information about the chemical state of selenium and zinc, narrow scan spectra of elemental lines were recorded from an analysis area of 0.5 x 0.5 mm^2 and with a pass energy of 23.5 eV. All spectra were charge referenced to the C 1s elemental line of adventitious hydrocarbon at 284.8 eV.

Spectra of selenium (Se $3p_{1/2}$, Se $3p_{3/2}$, Se $3d_{5/2}$ and Se $3d_{3/2}$) and zinc (Zn $2p_{3/2}$) were fitted using PHI MultiPak Version 9.4 (data analysis program). The background subtracted elemental lines (Shirley background) were fitted by applying a non-linear least-squares optimization procedure using Gaussian–Lorentzian sum functions. Oxidation states were identified by comparison with binding energies reported in the literature.

In order to assign unambiguously the obtained binding energies by XPS, reference selenium spectra from former studies were used for comparison. This survey highlighted that each oxidation state of Se has its specific binding energy, based on Se $3p_{3/2}$ and Se $3d_{5/2}$ peaks (Table S1). The binding energy difference between each oxidation state of selenium (−II, 0, IV and VI) is always higher than 1 eV for both Se $3p_{3/2}$ and Se $3d_{5/2}$ signals, i.e. clearly above the standard deviation of binding energies ($\pm\,0.2$ eV). This allows an accurate identification and discrimination between the different oxidation states of selenium.

Adsorption experiments to confirm absence of precipitation

To confirm the absence of precipitation in the control sorption experiments, a control at the same pH and zinc concentration as of the data point but without BioSeNPs was run. The initial zinc concentration of the control was measured before the 16 h incubation. After the incubation, to check for possible precipitation, the sample was centrifuged at 37,000g and the zinc concentration in the supernatant was measured. To check for any retention or release of zinc by the filter, supernatant after the centrifugation was filtered with a 0.45 µm syringe filter (cellulose acetate, Sigma Aldrich) and the zinc concentration was measured in the filtrate. All the determined zinc concentrations were always within 2% of standard deviation.

Figures

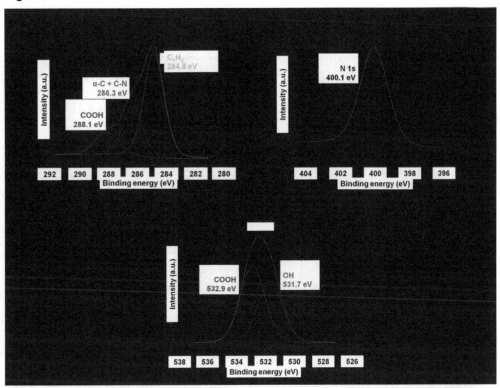

Figure S1. XPS spectra of BioSeNPs: (a) C 1s spectral lines suggesting the presence of hydrocarbon chains (C_xH_y), alpha-carbon (α-C) + C-N, and carboxylic acid (COOH groups; (b) N 1s spectral line suggesting the presence of nitrogen based compounds (amine or amide groups) and (c) O 1s spectral lines suggesting the presence of hydroxyl (OH) groups and carboxylate (COOH) groups.

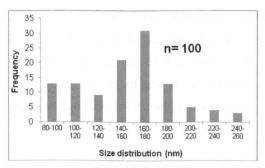

Figure S2. Size distribution (diameter) of BioSeNPs (n = 100).

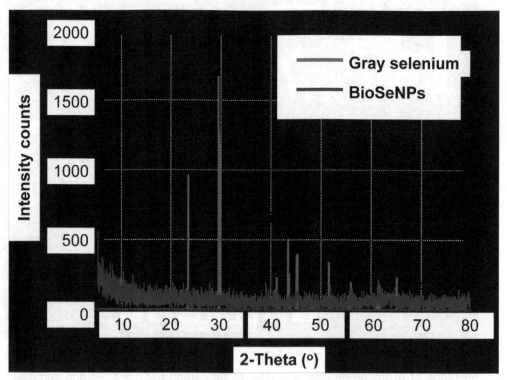

Figure S3. XRD pattern of gray trigonal selenium (reference) and BioSeNPs after purification.

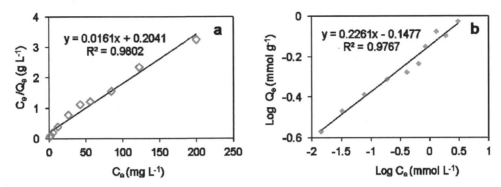

Figure S4. Adsorption isotherm data fitted with the (a) Langmuir and (b) Freundlich model.

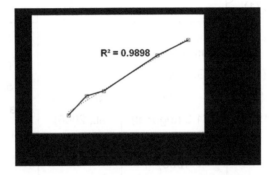

Figure S5. Moles of zinc ions adsorbed per mole of H$^+$ sorbed.

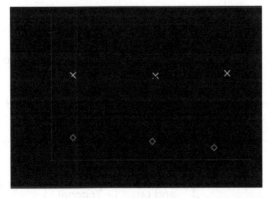

Figure S6. ζ-potential measurements of BioSeNPs (◇); BioSeNPs + calcium (×) and BioSeNPs + magnesium (−) at a background electrolyte concentration of 1 mM NaCl. Error bars indicate standard deviation from triplicate measurements.

Tables

Table S1. Binding energies (eV) of Se $3p_{3/2}$, Se $3d_{5/2}$ and Se 3d signal of Se reference compounds.

Compound	Se $3p_{3/2}$			Se $3d_{5/2}$	Se 3d
$Na_2Se^{VI}O_4$	164.6(0.1) (Swartz et al., 1971)	165.8(0.1) (Jordan, 2008)	–	59.4(0.1) (Jordan, 2008)	–
$Na_2Se^{IV}O_3$	164.1(0.1) (Swartz et al., 1971)	164.1 (Wagner, 1975)	164.1(0.1) (Jordan, 2008)	57.9(0.1) (Jordan, 2008)	–
Se(0)	161.9 (0.2) (metal) (Behl et al., 1980)	161.4 (amorphous)[a] (Guo and Lu, 1998) 161.3 (trigonal) (Guo and Lu, 1998)	161.6 (Sasaki et al., 2008)	Se powder 55.3 (Sasaki et al., 2008)	Amorphous 55.0 (Guo and Lu, 1998)
Na_2Se	–	–	158.4(0.1)[b] (Jordan, 2008)	–	52.7(0.1)[b] (Jordan, 2008)

[a] Se 3p signal

[b] a strong O 1s line indicating surface oxidation to Se(IV) was also observed

Table S2. Binding energies (eV) of Se 3d signal of Se(0) reported in the literature.

Se $3d_{5/2}$		Se 3d			
Grey crystalline electrodeposited 55.5 (Cannava et al., 2002)	Se powder 55.3 (Sasaki et al., 2008)	Amorphous 55.0 (Guo and Lu, 1998) Trigonal 55.0 (Guo and Lu,	Trigonal 54.9 (Xi et al., 2006)	Trigonal 54.32 (Yang et al., 2008)	Monoclinic 55.74 (Wang et al., 2010)

		1998)			

Table S3. Binding energies (eV) of Se 3p, Se 3d and Zn $2p_{3/2}$ signal of zinc compounds reported in the literature.

Compound	Se $3p_{3/2}$	Se 3d	Se $3d_{5/2}$	Zn $2p_{3/2}$
ZnSe (Swartz et al., 1971)	159.7	–	–	–
ZnSe (Islam et al., 1996)	–	54.2	–	1022.0
ZnSe (Sasaki et al., 2008)	161.0	–	54.9	–
ZnO (Ennaou et al., 1998)	–	–	–	1022.6
Zn(OH)$_2$ (NIST, 2012)	–	–	–	1021.8, 1022.7
ZnCO$_3$				1022.5

Table S4. Ionic radius, pauling electronegativity, standard reduction potential of Zn^{2+}, Ca^{2+} and Mg^{2+} ions.

Ion	Ionic radius (Å)	Pauling electronegativity (a.u.)	Std. reduction potential vs. NHE(V) $M^{2+} + 2e^- = M$
Zn^{2+}	0.74	1.65	−0.762
Ca^{2+}	0.99	1	−2.87

Mg^{2+}	0.65	1.31	−2.36
Fe^{2+}	0.57	1.83	-0.44

Table S5. ζ-potential values at different Zn/BioSeNPs ratio at pH range of 5.8-6.5.

Zn/BioSeNPs ratio	Qe-Zn (mg g^{-1})	ζ-potential (mV)
0	0	-36.8
0.0009	0.8	-31.2
0.0045	3.6	-28.6
0.0090	6.4	-28.1
0.0182	13.4	-22
0.0455	13.5	-18.6
0.1818	23	-13.5
0.3010	38.9	-11.9
0.4545	64.5	-10.1

References

Bahl, M. K.; Watson, R. L.; Irgolic, K. J., 1980. *L M M* Auger-spectra of selenium and some of its compounds. J. Chem. Phys. 72, 4069-4077.

Canava, B.; Vigneron, J.; Etcheberry, A.; Guillemoles, J. F.; Lincot, D.,2002. High resolution XPS studies of Se chemistry of a Cu(In, Ga)Se2 surface. Appl. Surf. Sci. 202, 8-14.

Ennaoui, A.; Weber, M.; Scheer, R.; Lewerenz, H. J.,1998. Chemical-bath ZnO buffer layer for CuInS$_2$ thin-film solar cells. Sol. Energy Mater. Sol. Cells 54, 277-286.

Guo, F. Q.; Lu, K., 1998. Microstructural evolution in melt-quenched amorphous Se during mechanical attrition. Phys. Rev. B 57, 10414-10420.

Islam, R.; Rao, D. R., 1996. X-ray photoelectron spectroscopy of Zn1-xCdxSe thin films. J. Electron Spectrosc. Relat. Phenom. 81, 69-77.

Jordan, N, 2008. Influence de l'acide silicique sur la rétention du selenium(IV) sur des oxydes de fer (in French). Ph.D Dissertation, Université de Nice-Sophia Antipolis, Nice

NIST X-ray Photoelectron Spectroscopy Database, 2012 Version 4.1 (National Institute of Standards and Technology, Gaithersburg); http://srdata.nist.gov/xps/.

Sasaki, K.; Blowes, D. W.; Ptacek, C. J., 2008. Spectroscopic study of precipitates formed during removal of selenium from mine drainage spiked with selenate using permeable reactive materials. Geochem. J. 42, 283-294.

Seah, M. P.; Gilmore, L. S.; Beamson, G., 1998. XPS: Binding energy calibration of electron spectrometers 5 - Re-evaluation of the reference energies. Surf. Interface Anal. 26, 642-649.

Swartz, W. E.; Wynne, K. J.; Hercules, D. M., 1971. X-ray photoelectron spectroscopic investigation of group of VI-A elements. Anal. Chem. 43, 1884-1887.

Wagner, C. D., 1975. Chemical-shifts of Auger lines, and Auger parameter. Faraday Discuss 60, 291-300.

Wang, T. T.; Yang, L. B.; Zhang, B. C.; Liu, J. H., 2010. Extracellular biosynthesis and transformation of selenium nanoparticles and application in H_2O_2 biosensor. Colloid Surf. B-Biointerfaces 80, 94-102.

Xi, G. C.; Xiong, K.; Zhao, Q. B.; Zhang, R.; Zhang, H. B.; Qian, Y. T., 2006. Nucleation-dissolution-recrystallization: A new growth mechanism for t-selenium nanotubes. Cryst. Growth Des. 6, 577-582.

Yang, L. B.; Shen, Y. H.; Xie, A. J.; Liang, J. J.; Zhang, B. C., 2008. Synthesis of Se nanoparticles by using TSA ion and its photocatalytic application for decolorization of cango red under UV irradiation. Mater. Res. Bull. 43, 572-582.

Appendix 3

This appendix was submitted as the experimental report form for:
Jain, R., Hullebusch, E.D. Van, Huguenot, D., Farges, F., Lens, P.N.L., 2014. Deciphering mechanism of interaction between biologically produced selenium nanoparticles and zinc metal ion. ESRF, Beamline BM-26, Experiment no. 26-01-991.

A) Overview:

Understanding the interaction of metals in general - Zn in this case - and biologically produced colloidal elemental selenium (BioSeNPs) is essential to determine the fate, mobility and toxicity of Se in the environment (Jain et al., 2014). To this point, this interaction was studied under different pH, ionic strength and Zn ion concentrations. The difference in ζ-potential measurements for BioSeNPs and BioSeNPs + Zn (-36 mV to -15 mV) suggests the electrostatic nature or covalent bond formation (can be either inner or outer sphere complex or both) interaction. X-ray photoelectron spectroscopy (XPS) analysis of Bio Se + Zn samples suggests the formation of an unidentified covalent Zn-Se phase (Jain et al., 2015). However, this interaction cannot be confirmed due to close proximity of binding energy of ZnO to ZnSe. Thus X-ray Absorption Spectroscopy (XAS) is required to fully understand this interaction. The EXAFS measurements carried out at Dubble beam line has allowed us, for the first time, to identify the first and second neighbours of Zn adsorbed on the surface of BioSe under different experimental conditions.

B) Data quality:

The measurements were successful and data recorded was of high quality. A variety of model compounds (those which were not measured during earlier experiments) and samples were measured at Zn-k edge. Even though the Zn concentration was relatively "low" (around 2000 ppm) in some samples, we were able to record the data successfully, thanks to the Ge-solid state detector. Also, we used a new graphical interface called as Generic Data Aquisition (GDA) installed at Dubble beamline for the first time and it worked quite well. To this end, we would like to acknowledge the help from Dr. Dip Banerjee and Dr. Alessandro Longo for their valuable support through out the beam time.

C) Status and progress of evaluation:

Primary data reduction has already been carried out. More detailed analysis and Feff ab-initio calculations and modeling will be carried out in the following months. The primary data analysis suggests to some interesting findings (please see section D for more details) and excellent signal/noise ratio related to excellent beam stability and detector sensitivity.

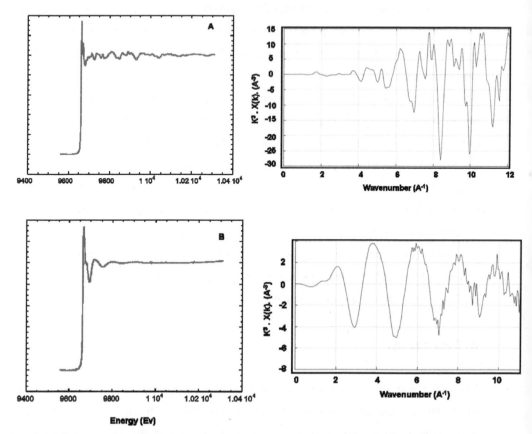

Figure A3.1. Zn K-edge data collected for (A) model compound with k^3 weighted EXAFS in transmission mode and (B) for a sample at high zinc loading along with k^3 weighted EXAFS in fluorescent mode at 50 K.

D) Results:

The primary data analysis already suggests intriguing results to be debvelopped incl. that Zn is adsorbed to BioSeNPs in different ways under different experimental conditions. In some cases, the $ZnCO_3$ precipitation is possible and in other cases, Zn is linked to oxygen like many organic compounds (Zn-acetate, Zn-lactate and so on). The results obtained at DUBBLE indicates the interaction of Zn is primarily to the polymeric substances attached to the surface of BioSeNPs. This understanding would help us to more closely predict the fate of BioSeNPs in the environment.

References:

Jain, R.; Gonzalez-Gil, G.; Singh, V., van Hullebusch, E.D., Farges, F.; Lens, P.N.L., 2014. Biogenic selenium nanoparticles, Production, characterization and challenges. In Kumar, A., Govil, J.N., Eds. Nanobiotechnology. Studium Press LLC, USA, pp. 361-390

Jain, R., Jordan, N., Schild, D., Hullebusch, E.D. Van, Weiss, S., Franzen, C., Hubner, R., Farges, F., Lens, P.N.L., 2015. Adsorption of zinc by biogenic elemental selenium nanoparticles. Chem. Eng. J. 260, 850–863.

Appendix 4

This appendix will be submitted as Supporting Information for

Jain, R., Matassa, S., Singh, S., Hullebusch, E.D. Van, Esposito, G., Lens, P.N.L., 2015. Reduction of selenite to elemental selenium nanoparticles by activated sludge under aerobic conditions.Process Biochem (*to be submitted*)

Transmission electron microscopy - Energy disperse X-ray spectroscopy

Samples for TEM were diluted in distilled water, deposited on a formvar-coated TEM grid and dried at ambient temperature. The analyses were performed using a JEOL 2100F (FEG) operating at 200 kV and equipped with a field emission gun, a high-resolution UHR pole piece, and a Gatan energy filter GIF 200. EDXS analysis were performed on the selected zone at 15 kV

DNA extraction and DGGE analysis

DNA was isolated using the FAST DNA SPIN kit from MP Biomedicals, USA. The DNA was isolated in accordance with the method given by Ahammad et al. (2013). The concentration of DNA in the isolated samples was 40 ng μL^{-1}. A nested PCR strategy was employed in which the first round PCR (Bio-Rad, C1000 Thermocycler, USA) was performed with primer set PRA46f and PRA1100r of amplicon size of 1054 bp. The following thermal cycling was used: Initial denaturation at 92°C for 3 min followed by 30 cycles of 92°C for 1 min, annealing at 55°C for 1 min, with a final elongation step at 72°C for 7 min. In the second round, the PCR products of first round were re-amplified with a set of universal primers PARCH340f-GC and PARCH519r with the following thermal cycling programme: Initial denaturation at 95°C for 3 min followed by 30 cycles of 92°C for 1 min, annealing at 55°C for 1 min, with a final elongation step at 72°C for 7 min.

In the first round of amplification the following recipe was used for making 25 μL PCR reaction mix. PCR master mix (Bioline, UK) 12.5 μL, H_2O 9.5 μL, forward primer PRA46f 0.5 μL, reverse primer PRA1100r 0.5μL, template DNA 2 μL. In the nested PCR, 50 μL reaction mix was prepared using 25 μL PCR master mix (Bioline, UK), 21 μL H_2O, 1.0 μL forward Primer (P340f), reverse Primer (P519r-GC) having added GC clams 1.0μL, template DNA 1μL (amplified DNA from the first round of PCR).

Primers UNIBACT341f-GC and UNIBACT518r were used for amplification of bacterial (V_3) 16S rRNA gene fragment. The following thermo cycler programme was used for amplification of bacterial V_3 region: 95°C for 3 min initial denaturation followed by 35 cycles of 95°C for 1 min, a touchdown protocol was used at annealing step with 65 to 55°C for 30 second with a decrement of (-0.5°C /step), elongation steps 72°C for 1 min and 72°C for 7 min for final elongation. The sizes of amplicon was checked by electrophoresis in 1.5% (wt./V) agarose gel stained with ethidium bromide. The list of primers used for PCR is provided in Table S3.

The denaturing gradient gel electrophoresis (DGGE) technique was used to estimate the community profile of the sludge samples. DGGE was performed in accordance with the protocol given by Muyzer et al. (1993). Bio-Rad D Code Universal Gene Mutation System (Bio-Rad Laboratories, Hercules, CA, USA) was used for running the gel. The PCR amplified products of the second round of the nested PCR were loaded on 8% polyacrylamide gels in 1% TAE (20 mmol/L Tris, 10 mM acetate and 0.5 mM EDTA pH 7.4 and a gradient of 45-60 % was maintained. The gel was run at 60 °C and 70 V for 16 h. Immediately after the gel electrophoresis, the plates were removed from the D-Code assembly and soaked in SYBR gold for 30 min for staining the gels. The stained gels were photographed in a Gel Documentation imaging system (Bio-Rad Laboratories, Hercules, CA, USA). Band pattern obtained were subjected to digital analysis. The intensities of the bands were analysed by Gel Doc XR+, Image LabTM 2.0 (Bio-Rad, Hercules, CA, USA).

Tables

Table S1. Operating parameters of continuous activated sludge reactor with sludge recycle

Parameter	Value
Influent flow (R1) [L h^{-1}]	0.125
Effluent flow (R2) [L h^{-1}]	0.250
Recirculation flow (R3) [L h^{-1}]	0.125
HRT [h]	8
Glucose (in terms of COD) [mg L^{-1}]	1000
OLR [gCOD L^{-1}d^{-1}]	3
Selenite [mM]	0.1
SeLR [mg Se L^{-1}d^{-1}]	~23.7

Table S2. List of primers used in the PCR for amplification of 16S rRNA gene

Name	Target group	Function	Sequences (5'-3')	References
PRA46f	Archaea	Forward	(C/T)TAAGCCATGC(G/A)AGT	(Øvreås et al., 1997)
PRA1100r		Reverse	(T/C)GGGTCTCGCTCGTT(G/A)CC	(Øvreås et al., 1997)
PARCH340fGC	Archaea, V3 region	Forward	CCCTACGGGG(C/T)GCA(G/C)CAG CGCCCGCCGCGCGCGGC GGGCGGGGCGGGGGCAC GGGGGG	(Øvreås et al., 1997)
PARCH519r		Reverse	TTACCGCGGC(G/T)GCTG	(Øvreås et al., 1997)
UNIBACT341fGC	Bacteria, V3 region	Forward	ACTCCTACGGGAGGCAGCAG CGCCCGCCGCGCGCGGC	[4]

			GGGCGGGGCGGGGGCAC GGGGGG	
UNIBACT518r		Reverse	ATTACCGCGGCTGCTGG	(Muyzer et al., 1993)

Figures

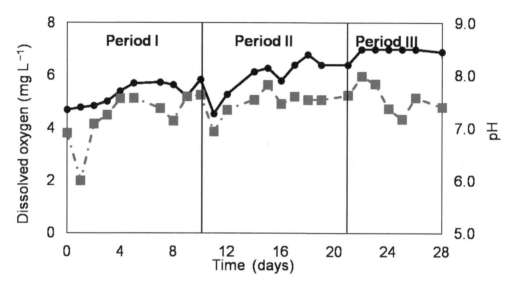

Figure S1. Dissolved oxygen concentration (left hand Y axis, ●) and pH (right hand Y axis, ■) in the continuous reactor for period I, II and III at TSS of 1300 mg L^{-1}

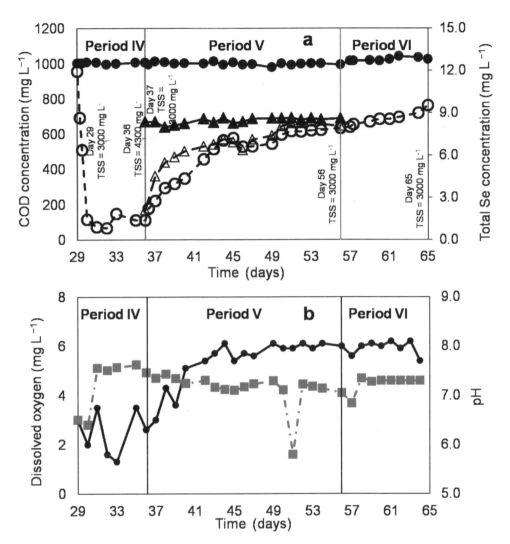

Figure S2. (a) Evolution of the COD (•, ○) and total selenium concentration (▲, △) in the influent (closed symbols) and effluent (open symbols) fed as selenite to a continuously aerated activated sludge reactor with complete sludge recycle at a TSS of 3000 mg L^{-1} and (b) Dissolved oxygen concentration (left hand Y axis, •) and pH (right hand axis Y axis, ■) in the continuous reactor for period IV, V and VI at TSS of 3000 mg L^{-1}

References

Ahammad, S.Z., Davenport, Read, L.F., Gomes, J., Sreekrishnan, T.R., Dolfing, J., 2013. Rational immobilization of methanogens in high cell density bioreactors. RSC Adv. 3, 774–781.

Lane,D.J., 1991. 16S/23S rRNA sequencing. In Stackebrandt. E and Goodfellow, M. Eds. Nucleic acid techniques in bacterial systematic, John Wiley & Sons , New york, pp 115-175

Muyzer, G., de Waal, E., Uitterlinden, A.G., 1993. Profiling of complex microbial populations by denaturing gradient gel electrophoresis analysis of polymerase chain. Appl. Environ. Microbiol. 59, 695–700.

Øvreås, L., Forney, L., Daae, F.L., Torsvik, V., 1997. Distribution of bacterioplankton in meromictic Lake Saelenvannet , as determined by denaturing gradient gel electrophoresis of PCR-amplified gene fragments coding for 16S rRNA. Appl. Environ. Microbiol. 63, 3367.

Curriculum vitae

Rohan Jain was born on 8th December, 1983 in Udaipur, India. He carried out his schooling in Central Academy in his hometown. He scored 1078 all India rank among 250,000 candidates to get admission into Indian Institute of Technology, Delhi, India for his bachelor and master degree. He pursued bachelor and master of technology in biochemical engineering and biotechnology. He was awarded "A" grade for his master thesis. Upon graduation, Rohan worked in a management consultancy company, Kinapse, catering the needs of life science industry. After working in the company for two year, Rohan moved to Paris to work at Institute of System and Synthetic Biology in the field of synthetic biology. He worked on the production of zinc fingers to regulate the promoters of *Escherichia coli*. In the beginning of 2011, Rohan got admitted into Erasmus Mundus Joint Doctorate Program on Environment Technologies for Contaminated Solids, Soils and Sediments (ETeCoS[3]). He carried out his PhD research at UNESCO-IHE and Université Paris-Est. His research was focused on deciphering the properties of biogenic elemental selenium nanoparticles.

Publications and conferences

Publications

Jain, R.; Gonzalez-Gil, G.; Singh, V., van Hullebusch, E.D., Farges, F.; Lens, P.N.L., 2014. Biogenic selenium nanoparticles, Production, characterization and challenges. In Kumar, A., Govil, J.N., Eds. Nanobiotechnology. Studium Press LLC, USA, pp. 361-390

Jain, R., Jordan, N., Weiss, S., Foerstendorf, H., Heim, K., Kacker, R., Hübner, R., Kramer, H., Hullebusch, E.D. Van, Farges, F., Lens, P.N.L., 2015. Extracellular polymeric substances (EPS) govern the surface charge of biogenic elemental selenium nanoparticles (BioSeNPs). Environ. Sci. Tech. 49, 1713-1720

Jain, R., Jordan, N., Schild, D., van Hullebusch, E.D., Weiss, S., Franzen, C., Hubner, R., Farges, F., Lens, P.N.L., 2015. Adsorption of zinc by biogenic elemental selenium nanoparticles. Chem. Eng. J. 260, 850–863.

Jain, R., Seder-Colomina, M. Jordan, N., Cosmidis, J., Hullebusch, E.D. Van, Weiss, S., Dessi, P., Farges, F., Lens, P.N.L., 2015. Entrapped elemental selenium nanoparticles increases settleablity and hydrophilicity of activated sludge. J. Hazad. Mat. 295, 193-200.

Jain, R., Pathak, A., Sreekrishnan, T. R., Dastidar, M.G., 2010. Autoheated thermophilic aerobic sludge digestion and metal bioleaching in a two-stage reactor system, J. Environ. Sci. 22, 230–236

Singh, V., **Jain, R.**, Dhar, P.K., 2014. Challenges and opportunities for synthetic biology in India, Current Synthetic and Systems Biology, Editorial, DOI: http://dx.doi.org/10.4172/2332-0737.1000e11

Singh, V., Chaudhary, D.K., Mani, I., **Jain, R.**, Mishra, B.N., 2013. Development of diagnostic and vaccine markers through cloning, expression, and regulation of putative virulence-protein-encoding genes of *Aeromonas hydrophila*, Journal of Microbiology 51, 275-282.

Conferences

Jain, R., Jordan, N., Weiss, S., Foerstendorf, H., Heim, K., Kacker, R., Hübner, R., Kramer, H., Hullebusch, E.D. Van, Farges, F., Lens, P.N.L., 2014. Extracellular polymeric substances govern the surface charge of biogenic elemental selenium nanoparticles. Paper presented at *Selen 2014 conference* in Karlsruhe, Germany.

Jain, R., Espinosa-Ortiz, E., Gonzalez-Gil, G., van Hullebusch, E.D., Farges, F., Lens, P.N.L., 2013. Microbial production of selenium nanoparticles. Paper presented at *Proceedings of the 3rd International Conference on Research Frontiers in Chalcogen Cycle Science & Technology* in Delft, The Netherlands.

Jain, R., Jordan, N., Schild, D., Hullebusch, E.D. Van, Weiss, S., Franzen, C., Hubner, R., Farges, F., Lens, P.N.L., 2015. Adsorption of zinc by biogenic elemental selenium nanoparticles. Paper presented at *Selen 2012 conference* in Karlsruhe, Germany.

Jain, R., Jordan, N., Weiss, S., Foerstendorf, H., Heim, K., Kacker, R., Hübner, R., Kramer, H., Hullebusch, E.D. Van, Farges, F., Lens, P.N.L., 2014. Extracellular polymeric substances govern the surface charge of biogenic elemental selenium nanoparticles. Poster presented at *International School on Crystallization* in Granada, Germany.

Jain, R., Hullebusch, E.D. Van, Farges, F., Lens, P.N.L., 2012. Biogenic production of elemental selenium nanoparticles. Poster presented at *EMBO Workshop on Microbial Sulfur Metabolism* in The Netherlands.

Netherlands Research School for the
Socio-Economic and Natural Sciences of the Environment

D I P L O M A

For specialised PhD training

The Netherlands Research School for the
Socio-Economic and Natural Sciences of the Environment
(SENSE) declares that

Rohan Jain

born on 8 December 1983 in Udaipur, India

has successfully fulfilled all requirements of the
Educational Programme of SENSE.

Delft, 19 December 2014

the Chairman of the SENSE board   the SENSE Director of Education

Prof. dr. Huub Rijnaarts   Dr. Ad van Dommelen

K O N I N K L I J K E   N E D E R L A N D S E
A K A D E M I E   V A N   W E T E N S C H A P P E N

The SENSE Research School declares that Mr Rohan Jain has successfully fulfilled all requirements of the Educational PhD Programme of SENSE with a work load of 50 EC, including the following activities:

SENSE PhD Courses

- Environmental Research in Context (2012)
- Research Context Activity: 'Contributing analysis to review IPCC chapter under coordination of the Netherlands Environmental Assessment Agency' (2013)
- Speciation and Bioavailability (2013)

Other PhD and Advanced MSc Courses

- Nanotechnology for water and wastewater treatment, UNESCO-IHE Delft (2011)
- Wastewater treatment technologies and modelling, University Paris-Est (2012)
- Summer school 'Contaminated sediments - characterisation and remediation', Delft, The Netherlands (2013)
- 4th Granada International School of Crystallisation, Laboratory of Crystallographic Studies, Granada, Spain (2014)

External training at a foreign research institute

- Summer school 'Contaminated Soils: From characterization to remediation' University Paris-Est, France (2012)
- Summer School 'Biological Treatment of Solid Waste', Cassino and Gaeta, Italy (2014)
- X-ray Absorption Spectroscopy, European Synchrotron Radiation Facility (ESRF), Grenoble, France (2014)
- Gaussian 9.0 Software, Helmholtz-Zentrum Dresden-Rossendorf (HZDR), Dresden, Germany (2014)

Management and Didactic Skills Training

- Supervision of three MSc students

Oral Presentations

- *Adsorption of zinc by biogenic elemental selenium nanoparticles.* Selen2012 Workshop - 'Selenium in geological, hydrological and biological systems', 8-9 October 2012, Karlsruhe, Germany
- *Microbial synthesis of selenium nanoparticles.* 3rd International Conference on Research Frontiers in Chalcogen Cycle Science & Technology, 27-28 May 2013, Delft, The Netherlands
- *Extracellular polymeric substances govern surface charge of biogenic elemental selenium nanoparticles.* Selen2014 Workshop - 'Selenium in geological, hydrological and biological systems', 13-14 October 2014, Karlsruhe, Germany

SENSE Coordinator PhD Education

Dr. ing. Monique Gulickx